LONGMANS' CIVIL ENGINEERING SERIES

RAILWAY CONSTRUCTION

BY

WILLIAM HEMINGWAY MILLS, M.Inst.C.E.

PAST PRESIDENT OF THE INSTITUTION OF CIVIL ENGINEERS OF IRELAND,
AND
ENGINEER-IN-CHIEF OF THE GREAT NORTHERN
RAILWAY OF IRELAND

WITH ILLUSTRATIONS

FOURTH IMPRESSION

LONGMANS, GREEN, AND CO.
39 PATERNOSTER ROW, LONDON
NEW YORK, BOMBAY, AND CALCUTTA

1910

All rights reserved

PREFACE

The construction and maintenance of a railway calls for the application of so many branches of engineering that several volumes would be required to do ample justice to a subject so comprehensive and ever-extending. To avoid attempting so wide a range, the object of the following pages has been to describe briefly some of the recognized leading features which regulate railway construction, and to assist the explanation with sketches of works selected from actual practice.

Where the number of existing good examples is legion, it is somewhat difficult to make a choice for illustration, and the course adopted has been to select such samples of structures as appear best to elucidate in a simple manner the different types of work under consideration.

In the drawings and diagrams many important minor details are necessarily omitted, partly to avoid complexity, but principally to leave more prominent the leading features of the particular piece of work referred to in the description. Some of the sketches of the large span bridges and large span roofs are only shown in outline; but, as their principal dimensions are given, a general idea can be obtained of their actual proportions.

No allusion is made to the requisite strengths of the various structures described, nor to the necessary dimensions of the materials used in their construction, as this would necessitate the introduction of a vast amount of mathematical formulæ which does not come under the province of the object in view, and which the engineer has already at command from his training and works of reference.

Neither is any mention made as to the probable cost of the different works of construction, as these must always vary to a very large extent, according to the locality, facility of supply, and current prices of materials.

Every railway scheme which is the outcome of public enterprise has its commercial aspect and influence. The large sums to be invested in its construction are expected to yield permanent and increasing returns, and this desirable end can only be attained where there is thorough efficiency in works and equipment, and a full compliance with those national regulations which control matters connected with public safety. The correct dealing with the technical requirements and structural features of the undertaking must always precede all other considerations, as the constituted authorities will exact a proper fulfilment of all the statutory obligations, regardless of the prospective remuneration to the promoters. A stroke of the pen may change a train-service, or alter the rates and tariffs, but a modification in the works of construction arising out of errors or oversight, would entail a heavy expenditure and tedious delay. The essential point of every railway undertaking must be its suitability and completeness in every respect for the duty for which it is intended.

Notes of what has been done are always valuable for consideration and comparison, and that the following brief description and sketches may be found useful for reference, is the earnest wish of the writer.

W. H. MILLS,
M.Inst.C.E.

CONTENTS

PAGE

CHAPTER I.

Location of a line of railway—Government regulations—Questions for consideration in connection with gauge, gradients, and curves 7

CHAPTER II.

Works of construction: Earthworks, Culverts, Bridges, Foundations, Screw piles, Cylinders, Caissons, Retaining walls, and Tunnels 63

CHAPTER III.

Permanent way—Rails—Sleepers—Fastenings—and Permanent-way laying 179

CHAPTER IV.

Stations: Station Buildings, Roofs, Lines, and Sidings 271

CHAPTER V.

Sorting-sidings—Turn-tables—Traversers—Water-Tanks and Water-Columns 301

Comparative Weights of some Types of Modern Locomotives   319

CHAPTER VII.

Signals—Interlocking—Block Telegraph and Electric Train Staff Instruments   327

CHAPTER VIII.

Railways of different ranks—Progressive improvements—Growing tendency for increased speeds, with corresponding increase in weight of permanent way and rolling-stock—Electricity as a motive-power   361

[Contents]

CHAPTER I.

Location of a line of railway—Government regulations—Questions for consideration in connection with gauge, gradients, and curves.

Location.—The locating of a line of railway, or the determination of its exact route, is influenced by many circumstances. In a rich country, with thickly populated districts and large industrial enterprises, there are towns to be served, manufacturing centres to be accommodated, and harbours to be brought into connection; while, at the same time, there may be important estates which must be avoided and private properties which must not be entered. Each point will present its own individual claim for consideration when selecting the route which promises the greatest amount of public convenience and commercial success.

In new countries—in our colonies, and especially out in the far west of Canada and the United States—railways have to be laid out in almost uninhabited districts, where there is but little population or commerce to serve, and where the principal object is to obtain the best and most direct route through the vast territories, leaving colonists and settlers to choose afterwards the most convenient sites for towns and villages. Untrammelled by the network of public and private roads and properties which are met with at home, it might appear that the locating of such a line would be comparatively light; but even in such countries, which at first sight seem to present unlimited freedom for selecting a route, much can be done, and should be done, by taking a course through those plains and districts which possess the best natural resources for future agricultural, manufacturing, or mineral development.

In addition to the motives of convenience and policy, the route of every line of railway must be influenced by the natural features of the country—the mountains, valleys, and rivers. These physical obstacles are in some cases on such an enormous scale as to compel long detours in the formation of a more suitable opening; and in others, although the difficulties are not insurmountable, they may involve works of great magnitude and expense.

In a comparatively rich country, with a prospect of large and remunerative traffic, a succession of heavy works, bridges, and tunnels may be admissible and expedient; but in new countries economy of outlay has to be considered, and costly works avoided as much as possible.

Every one of the heavy works on a line, whether lofty bridges, long viaducts, or costly tunnels, not only enormously increase the original expenditure of the undertaking, but also entail large annual outlay in the necessary constant supervision and maintenance.

Each particular scheme will have to be discussed on its own individual merits. The heavy, high-speed passenger traffic line will suggest light gradients and easy curves, while on secondary lines and in thinly populated districts it may be prudent, for the sake of economy, to introduce sharper curves and heavier gradients. Even in the latter case, and especially in new countries, it is well to keep in view the future possibilities of the undertaking. The steeper the gradients, the greater the cost and time in working the traffic, and if there is every probability of early and large development, the prospective increase may warrant an additional outlay in the original construction.

Large, open plains and wide valleys of important rivers generally afford ample latitude for the selection of a suitable route, and, by taking advantage of the gradations of altitude, a favourable course may be adopted without incurring excessive gradients. When traversing moderately hilly districts, some low ridge or opening may be found, which may form a pass from the one side to the other, and the line may be laid out for a long distance to lead gradually up to the highest point. But when a route has to be laid out over some of those lofty mountain ranges which are met with abroad, the locating of a suitable line, or of any line, becomes particularly intricate and difficult. A comparatively low ridge may be found possessing features in favour of the project, but the question will be how to reach that point. The nearer the summit of these high mountains, the more precipitous the sides; no one slope can be found sufficiently long and uniform to permit a practical direct ascent, and the only way out of the difficulty is to make a series of detours along the various spurs of the mountains to gain length to overcome the height. Each detour has to be the subject of most careful study. Forming part of a long series of ascending gradients, it has to follow the winding of the mountain-side, must be laid out to be always gaining in height, and will comprise important works, many of them of considerable extent, necessary for protection against the floods and atmospherical changes of the locality.

In these higher altitudes nature is met with on the grandest and most rugged scale. Deep gorges, wide ravines, and almost perpendicular rocks form the pathway along which the line must be carried, and the skill of the engineer is taxed to the utmost to select a course which shall comprise a minimum of the works of magnitude. Mile after mile of line must be laid out in almost inaccessible places, loose or broken rocks must be avoided, a firm foundation must be obtained at all points skirting high ledges, and ample provision must be made for those mountain torrents which rise so suddenly, and are liable to sweep away all before them.

Many grand examples of these detour lines are in existence in different parts of the world, and the traveller passing over them can realize the difficulties that had to be encountered, and the masterly manner in which they have been overcome.

Before proceeding to carry out the works of any line of railway, it is necessary to prepare a complete plan and section of the line, showing the route to be followed and the position of the various curves, gradients, and principal works. Within certain limits, the course of the line may have to be slightly modified as the work proceeds, in consequence of ground turning out unfavourable, river-crossings treacherous, or of sites involving so many contingent alterations that it is found better to avoid them altogether. The route should, however, be so carefully studied out before completing the final plan and section, as to leave only minor deviations of line and level to be dealt with in the actual carrying out of the work.

The promoters of lines in the United Kingdom obtain valuable assistance from the ordnance maps, which give full and reliable information regarding the position of all roads, rivers, and boundaries of counties, parishes, and townlands. In many parts abroad local maps are scarce, and not always accurate, and engineers have to depend principally on their own surveys, and rely upon the resident local authorities for any particulars as to divisions of territory. On some of our great colonial plains, and out in the far west of America, a line may be laid out for miles without a single landmark to localize it on a plan; but careful setting out, and the relative levels of the ground and gradients, as shown on the section, will always indicate the correct position of any portion of the work.

Both at home and abroad complete plans and sections of any proposed railway must be deposited with the proper Government authorities, and must be approved and sanctioned by them before permission can be obtained to proceed with the works.

The regulations regarding the scale and general arrangement of these plans and sections vary in different countries, and are subject to modification from time to time.

Each country has its own special enactments relative to the method of dealing with roads, rivers, streams, and public and private property proposed to be interfered with in the construction of any line, and a knowledge of these is absolutely necessary for the promoters of any new scheme, inasmuch as some of the requirements may, in certain instances, influence the precise route to be selected.

The English Government has passed several Acts of Parliament setting forth the general conditions which must be complied with in the construction of any railway in the United Kingdom. These conditions, or standing orders, relate both to the acquirement of land and property, the size and description of works for public or private accommodation, and the inspection and official approval of the undertaking when completed. These fixed regulations are alike valuable to the promoters and to the public; the former are informed of the principal points with which the scheme must conform, and the latter know the limit of their legal demands.

No line of railway, or extension of any railway, will obtain Parliamentary sanction unless it can be satisfactorily proved in the outset, that its construction would be of public advantage. This point is of paramount importance, and due weight must be given to it when preparing to refute the evidence of opponents to the scheme.

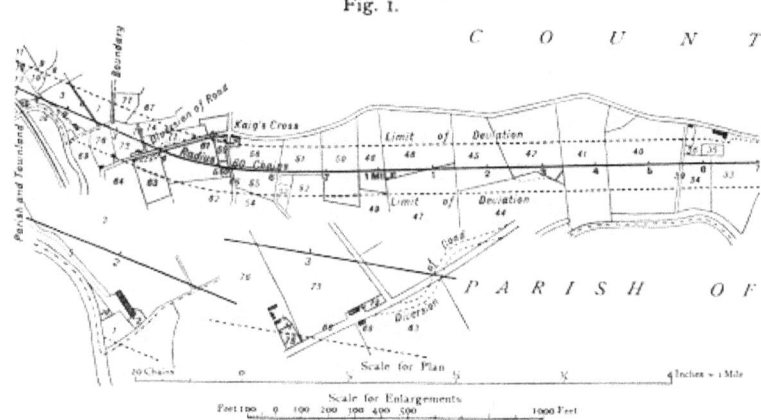

Fig. 1.

When conceding the right to make any railway, Parliament grants with it the power to purchase lands or property compulsorily, or by agreement, to change and divert roads and streams in the manner shown on the deposited plans, and to construct all necessary bridges and works in accordance with the standing orders, or such modifications of them as may be approved by the Board of Trade.

The standing orders, or Government regulations, are very comprehensive, and include much detailed information on all questions likely to arise. The following brief summary of some of the principal orders relating to deposited plans, and works of construction, will be found useful for reference.

Extract from Government Standing Orders and Regulations.—All plans and sections relative to proposed new railways must be lodged with the constituted Government Authorities on or before November 30.

Every deposited plan must be drawn to a scale of not less than four inches to a mile, and must describe the centre line, or situation of the work (no alternative line being allowed), and must show all lands, gardens, or buildings within the limits of deviation, each one being numbered with a reference number, and where powers to make lateral deviations are applied for, the limits of such deviation must be marked on the plan.

Unless the whole of such plan be drawn to a scale of not less than 400 feet to an inch, an enlarged plan must be drawn to that scale of every building and garden within the limits of deviation.

The Railway Clauses Act limits the extent of deviation to 100 yards on each side of the centre line in the country, and 10 yards on each side of the centre line in towns or villages.

The distances must be marked on the plan in miles and furlongs from one of the termini.

The radius of every curve not exceeding one mile must be marked on the plan in furlongs and chains.

In tunnels the centre line must be dotted, but no work must be shown as tunnelling, in the making of which it is necessary to cut through, or remove the surface soil. If it is intended to divert or alter any public road, navigable river, canal, or railway, the course and extent of such diversion, etc., shall be marked on the plan.

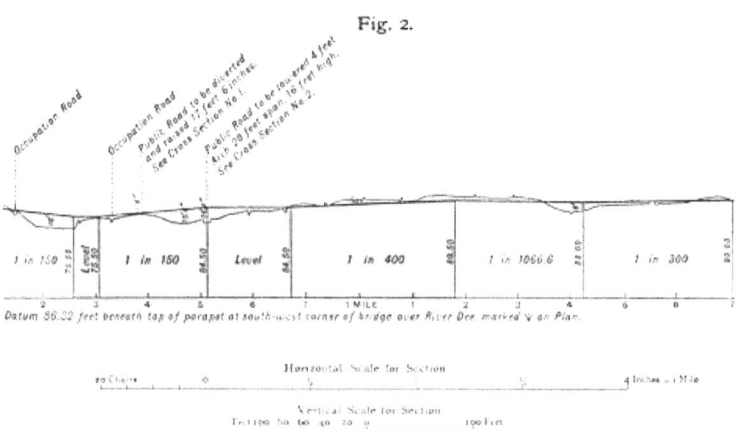

Fig. 2.

When a railway is to form a junction with an existing railway, the course of such existing railway must be shown on the plan for a distance of 800 yards on each side of the proposed junction. In the case of Bills for constructing subways, the plans and sections must indicate the height and width of such subway, and the nature of the approaches by which it is proposed to afford access to such subway.

The Book of Reference must contain the names of all owners, lessees, and occupiers of all lands and houses of every parish within the limits of deviation.

The numbers on the Book of Reference must correspond with the numbers on the plan, and opposite to each number must be entered a brief description of the property, whether field, garden, house, road, railway, or river. It is intended that the plan and Book of Reference together, shall afford ample information to enable all parties interested to ascertain to what extent their property will be affected by the proposed undertaking.

The section must be drawn to the same horizontal scale as the plan, and to a vertical scale of not less than 100 feet to an inch, and must show the level of the ground, the level of the proposed work, the height of every embankment, the depth of every cutting, and a horizontal datum line which shall be referred to some fixed point, near one of the termini.

In every section of a railway, the line of railway marked thereon must correspond with the upper surface of the rails.

Distances on the datum line must be marked in miles and furlongs to correspond with those on the plan; a vertical measure from the datum line to the line of the railway must be marked in feet and decimals at the commencement and termination of the railway, and at each change of gradient, and the rate of inclination between such vertical measures must also be marked.

Wherever the line of railway crosses any public carriage road, navigable river, canal, or railway, the height of the railway over, or depth beneath the surface thereof, and the height and span of every arch by which the railway will be carried over the same, must be marked in figures.

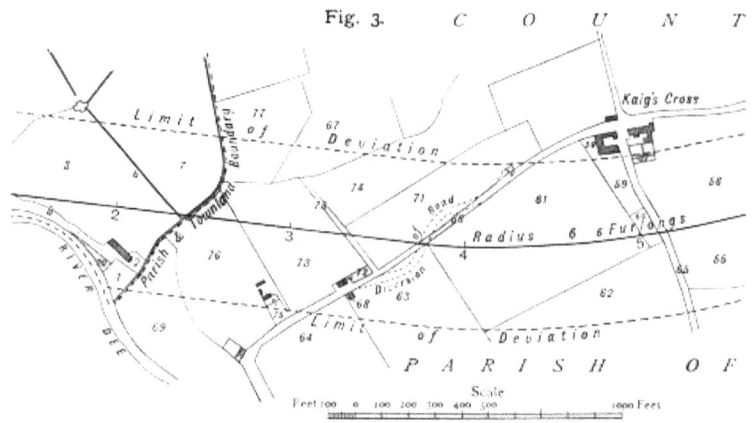

Fig. 3.

In the case of a public road level crossing, it must be described on the section, and it must also be stated if such level will be unaltered. If any alteration be intended in the level of any canal, public road, or railway which will be crossed by the intended line of railway, the same must be stated on the section and cross-sections to a horizontal scale of not less than 330 feet to an inch, and a vertical scale of not less than 40 feet to an inch must be added, which must show the present surface of such road, canal, etc., and the intended surface thereof when altered, and the greatest of the present and intended rates of inclination marked in figures, such cross-sections to extend 200 yards on each side of the centre line of railway.

Wherever the height of any embankment, or depth of any cutting, shall exceed 5 feet, the extreme height over or depth beneath the surface of the ground must be marked in figures upon the section.

All tunnels and viaducts must be shown on the section.

At a junction with an existing railway, the gradient of such existing railway must be shown on the section on the same scale as the general section for a distance of 800 yards on each side of the point of junction.

Where the level of any turnpike or public road has to be altered in making any railway, the gradient of any altered road need not be better than the mean inclination of the existing road within a distance of 250 yards of the point of crossing the railway; but where the existing roads have easy gradients, then the gradients of the altered roads, whether carried over, or under, or on the level with the railway, must not be steeper than 1 in 30 for a turnpike road, 1 in 20 for a public carriage road, 1 in 16 for a private or occupation road.

A good and sufficient fence, 4 feet high at least, shall be made on each side of every bridge, and fences 3 feet high on the approaches.

The application to cross any public road on the level must be reported upon by one of the officers of the Board of Trade, and special permission for the work must be embodied in the Act.

Not more than 20 houses of the labouring classes may be purchased in any city or parish in England, Scotland, and Wales, or more than 10 such houses in Ireland, until approval has been obtained to a scheme for building such houses in lieu thereof as the authorities may deem necessary.

Every bridge (unless specially authorized to be otherwise) must conform with the following regulations:— A bridge over a turnpike road must have a clear span of 35 feet on the square between the abutments, with a headway, or height, of 16 feet for a width of 12 feet, as shown on Fig. 12.

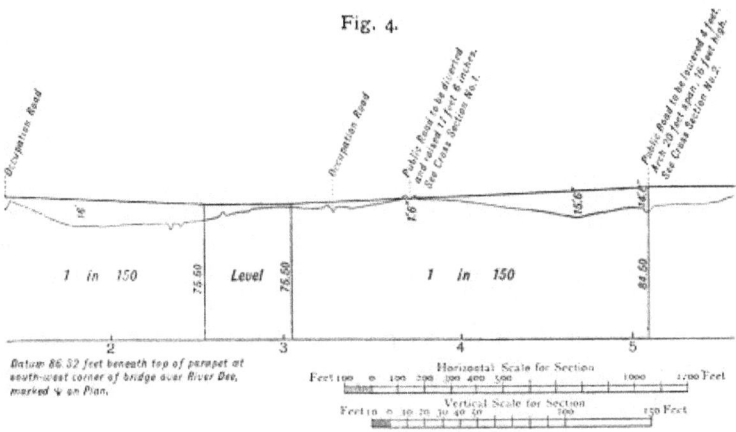

A bridge over a public road must have a clear span of 25 feet on the square between the abutments, with a headway of 15 feet for a width of 10 feet, as shown on Fig. 13.

A bridge over a private or occupation road must have a clear span of 12 feet on the square between the abutments, with a headway of 14 feet for a width of 9 feet, as shown on Fig. 14.

Road bridges over the railway must have the same clear width between the parapets, measured on the square, as the widths prescribed for road bridges under the railway, or 35 feet for a turnpike road, 25 feet for a public road, and 12 feet for private or occupation road.

It is not compulsory, however, to construct the public road bridges over or under the railway of a greater width than the average available width of the existing roads within 50 yards of the point of crossing the railway, but in no case must a bridge have a less width than 20 feet. Should the narrow roads be widened at any future time, the railway company will be under the obligation to widen the bridges at their own expense to the extent of the statutory widths of 35 feet for a turnpike road, and 25 feet for a public road.

Suitable accommodation works in the form of bridges, level crossings, gates, or other works, must be provided for the owners, or occupiers of lands, or properties intersected or affected by the construction of the railway; or payments may be made by agreement instead of accommodation works. All questions, or differences between the Railway Company, and the owners or occupiers of property affected, will be decided by the authorities duly appointed by the Government for the purpose.

In constructing the railway, the Parliamentary plans and sections may be deviated from to the following extent:—

The centre line may be deviated anywhere within the limits of deviation (100 yards on each side of the centre line in country, and 10 yards each side in towns, or villages).

Curves may be sharpened up to half a mile radius, and further, if authorized by the Board of Trade.

A tunnel may be made instead of a cutting, and a viaduct instead of an embankment, if authorized by the Board of Trade.

The levels may be deviated from to the extent of 5 feet in the country, and 2 feet in a town, or village, and various authorities have power to consent to further deviations.

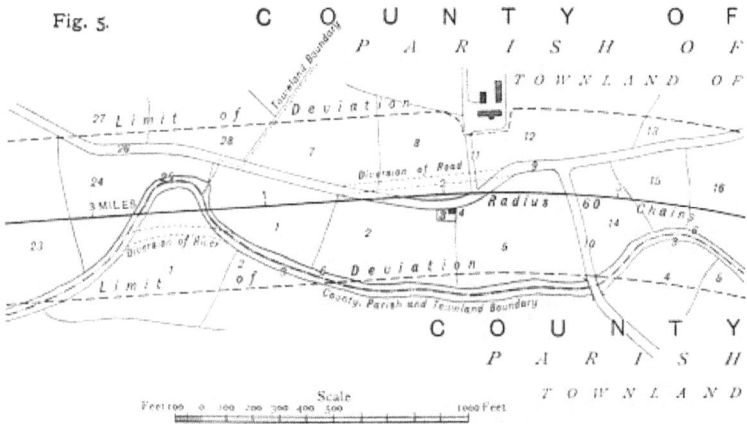

Fig. 5.

Gradients may be diminished to any extent, gradients flatter than 1 in 100 may be made steeper to the extent of 10 feet in a mile, and gradients steeper than 1 in 100 may be made steeper to the extent of 3 feet in a mile, or to such further extent as may be authorized by the Board of Trade.

Suitable fences must be erected on each side of the line, to separate the land taken for the use of the railway from the adjoining lands not taken, and to protect such lands from trespass, or the cattle of the owners, or occupiers thereof from straying on to the railway.

In addition to the Parliamentary plans, and sections, and Book of Reference, an estimate of the cost of each separate line, or branch, must be prepared as near to the following form as circumstances will permit.

[Transcriber's Note: Searchable text of the following form can be found at the end of the chapter.]

ESTIMATE OF THE PROPOSED (RAILWAY).
Line No. _____

Length of Line:	Miles.	F.	Chs.	Whether single or double.

	Cubic yards.	Price per yard.	£	s.	d.	£	s.	d.
Earthworks :—								
Cuttings—Rock...								
Soft soil ...								
Roads ...								
Total ...								
Embankments, including roads, ____ cubic yards	...	...						
Bridges, public roads—number	...	...	...	...				
Accommodation bridges and works ...	...	...	...	...				
Viaducts	...	...	...	...				
Culverts and drains	...	...	...	...				
Metallings of roads and level crossings	...	...						
Gatekeepers' houses at level crossings	...	...						
Permanent way, including fencing :—								

Miles.	F.	Chs.	Cost per mile.		
			£	s.	d.
—	—	— at			

Permanent way for sidings, and cost of junctions	...	...
Stations	...	...
Contingencies ____ per cent.	...	...	...	...

Land and buildings	A.	R.	P.	...	...
	—	—	—		

Total ... £

Dated this _____ day of _____ 18___

Witness : _____

Engineer.

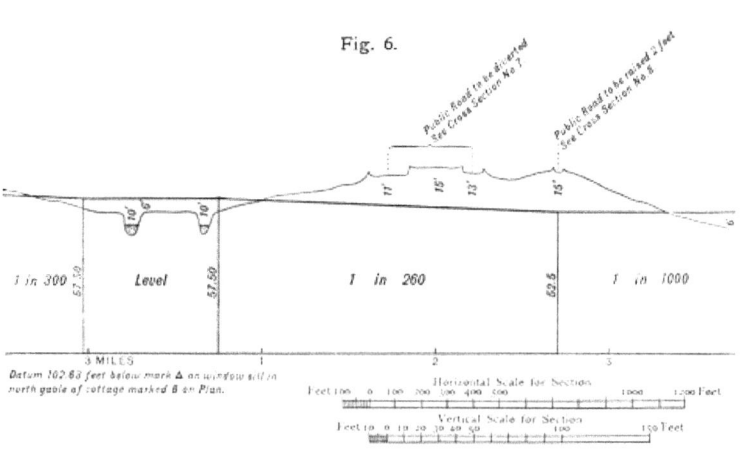

Fig. 6.

The same details for each branch, and general summary of total cost.

Every Railway Bill must be read twice, both in the House of

Commons and in the House of Lords. A committee, duly appointed for each House, must report upon it, and if the reports from such committees be favourable, the Bill will be read a third time, and passed.

When it has passed both Houses, the Bill receives the Royal Assent, and becomes law.

The minimum scale of four inches to a mile for the plans is so very small that it is rarely, if ever, adopted. It would necessitate enlarged plans of so many portions to show clearly the property or buildings inside the limits of deviation, that in practice it is found expedient to make the plans to a much larger scale.

Figs. 1 and 2 show a small portion of a Parliamentary plan and section drawn to the minimum scale allowed, with an enlargement of a small part to distinguish the houses clearly.

Figs. 3 and 4 show a part of the same plan and section drawn to a scale of 400 feet to an inch, a scale which is very frequently adopted, and is sufficiently large to distinguish the buildings and small plots, except in closely populated districts. This scale also gives ample room for reference numbers.

The Parliamentary plans and sections must be accurate in delineation, levels, and description. All property within the prescribed limits of deviation must be clearly shown, and the numbers and description on the plans and book of reference must be concise and complete, to enable the owners to ascertain to what extent they will be affected. In every place where it is proposed to interfere with any public highway, street, footpath, river or canal, the manner of such proposed alteration must be shown and described on both plan and section. The commencement and termination of every tunnel must be correctly indicated, and the length given on both plan and section. An omission of any of the above requirements might prove very detrimental to the scheme, and possibly result in the Bill being thrown out of Parliament for non-compliance with standing orders.

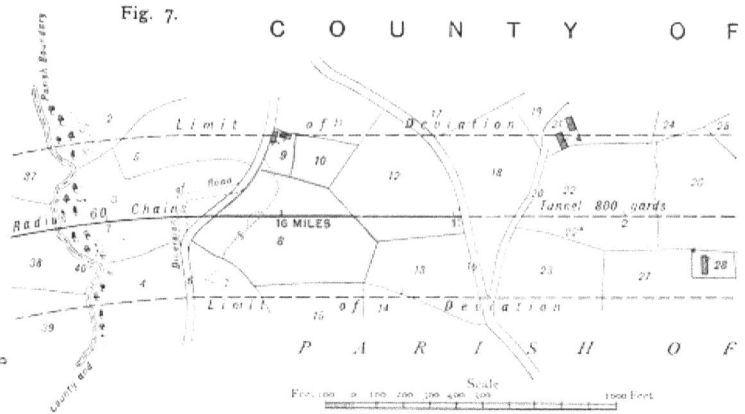

Fig. 7.

In carrying out the works the constructors have power to deviate the centre line either to the one side or the other, provided that such deviation will permit of the boundary of the works, or property to be acquired, to come within the limits of deviation or property referenced, and they may also vary the levels of the line to the extent prescribed in the standing orders.

Figs. 5 and 6 are parts of a Parliamentary plan and section showing alteration of a public road with an overline bridge—also a diversion of a small river to avoid two river bridges.

Figs. 7 and 8 are parts of a Parliamentary plan and section showing a public road diverted and carried under the railway.

A stipulated time is fixed in the Bill for the purchase of the property and construction of the line, and if this time be exceeded before the completion of the works, it will be necessary to obtain further Parliamentary powers for an extension of time.

Every new railway, or extension of railway, in the United Kingdom, must be inspected, and certified, by one of the inspecting officers of the Board of Trade, previous to Government sanction being granted for its opening as a passenger line.

To facilitate these inspections, and as a guide both to their own inspecting officers and the engineers in charge of the construction, the Board of Trade have issued a list of the principal requirements in connection with all new lines.

The following is a copy of the list so far as relates to works of construction and signals:—

Requirements of the Board of Trade.—1. The requisite apparatus for providing by means of the block telegraph system an adequate interval of space between following trains, and, in the case of junctions, between converging or crossing trains. In the case of single lines worked by one engine under steam (or two or more coupled together) carrying a staff, no such apparatus will be required.

2. Home-signals and distant-signals for each direction to be fixed at stations and junctions, with extra signals for such dock, or bay lines, as are used either for the arrival, or for the departure of trains, and starting-signals for each direction, at all passenger stations which are also block posts. On passenger lines all cross-over roads and all connections for goods, or mineral lines, and sidings to be protected by home and distant signals, and as a rule at all important running junctions a separate distant-signal to be provided in connection with each home-signal.

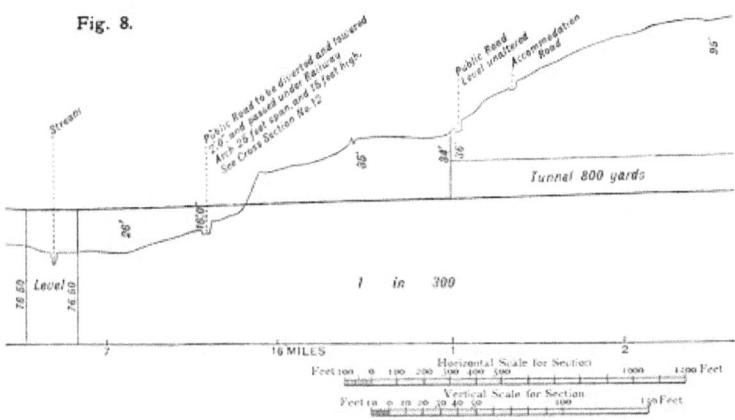

Fig. 8.

Signals may be dispensed with on single lines under the following conditions:—

> (a) *At all stations and siding connections upon a line worked by one engine only (or two engines coupled together), carrying a staff, and when all points are locked by such staff.*

(*b*) *At any intermediate siding connection upon a line worked under the train staff and ticket system, or under the electric staff or tablet system, where the points are locked by the staff or tablet.*

(*c*) *At intermediate stations, which are not staff or tablet stations, upon a line worked under the electric staff or tablet system: Sidings, if any, being locked as in (b).*

3. The signals at junctions to be on separate posts, or on brackets; and the signals at stations, when there is more than one arm on one side of a post, to be made to apply—the first, or upper arm, to the line on the left, the second arm to the line next in order from the left, and so on; but in cases where the main, or more important line, is not the one on the left, separate signal-posts to be provided, or the arms to be on brackets. Distant-signals to be distinguished by notches cut out of the ends of the arms, and to be controlled by home or starting signals for the same direction when on the same post. A distant-signal arm must not be placed above a home or starting signal arm on the same post for trains going in the same direction.

In the case of sidings, a low short arm and a small signal light, distinguishable from the arms or lights for the passenger lines, may be employed, but in such cases disc signals are, as a rule, preferable.

Every signal arm to be so weighted as to fly to and remain at danger on the breaking at any point of the connection between the arm and the lever working it.

4. On new lines worked independently, the front signal lights to be green for "all right," and red for "danger;" the back lights (visible only when the signals are at "danger") to be white.

This requirement not to be obligatory in the case of new lines run over by trains of other companies using a different system of lights.

5. Facing points to be avoided as far as possible, but when they cannot be dispensed with they must be placed as near as practicable to the levers by which they are worked or bolted. The limit of distance from levers working points to be 180 yards in the case of facing points, and 300 yards in the case of trailing points on the main line, or safety points of sidings.

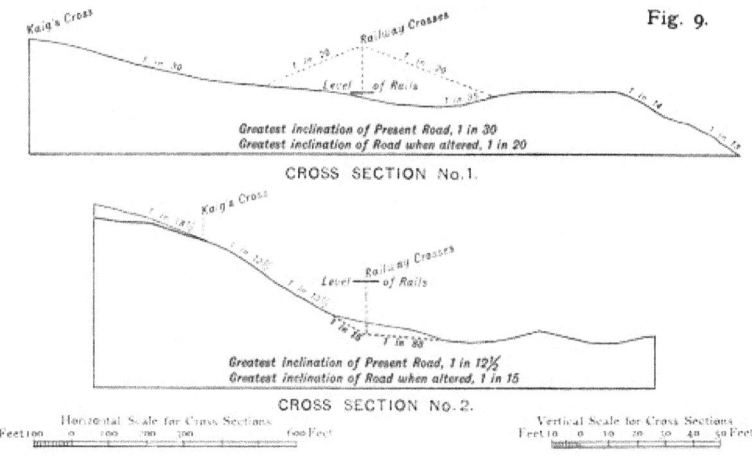

Fig. 9.

CROSS SECTION No. 1.

CROSS SECTION No. 2.

In order to ensure that the points are in their proper position before the signals are lowered, and to prevent the signalman from shifting them while a train is passing over them, all facing points must be fitted with facing-point locks and locking-bars, and with means for detecting any failure in the connections between the signal-cabin and points. The length of the locking-bars to exceed the greatest wheel-base between any two pairs of wheels of the vehicles in use on the line, and the stock rails to be tied to gauge with iron or steel ties. All points, whether facing or trailing, to be worked or bolted by rods, and not by wires, and to be fitted with double connecting-rods.

6. The levers by which points and signals are worked to be interlocked and, as a rule, brought close together, into the position most convenient for the person working them, in a signal-cabin or on a properly constructed stage. The signal-cabin to be commodious, and to be supplied with a clock, and with a separate block instrument for signalling trains on each line of rails. The point-levers and signal-levers to be so placed in the cabin that the signalman when working them shall have the best possible view of the railway, and the cabin itself to be so situated as to enable the signalman to see the arms and the lights of the signals and the working of the points. The back lights of the signal lamps to be made as small as possible, having regard to efficiency, and when the front lights are visible to the signalman in his cabin no back lights to be provided. The fixed lights in the signal-cabin to be screened off, so as not to be mistakable for the signals

exhibited to control the running of trains. If, from any unavoidable cause, the arm and light of any signal cannot be seen by the signalman they must, as a rule, be repeated in the cabin.

7. The interlocking to be so arranged that the signalman shall be unable to lower a signal for the approach of a train until after he has set the points in the proper position for it to pass; that it shall not be possible for him to exhibit at the same moment any two signals that can lead to a collision between two trains; and that, after having lowered the signals to allow a train to pass, he shall not be able to move any points connected with, or leading to, the line on which the train is moving. Points also, if possible, to be so interlocked as to avoid the risk of a collision.

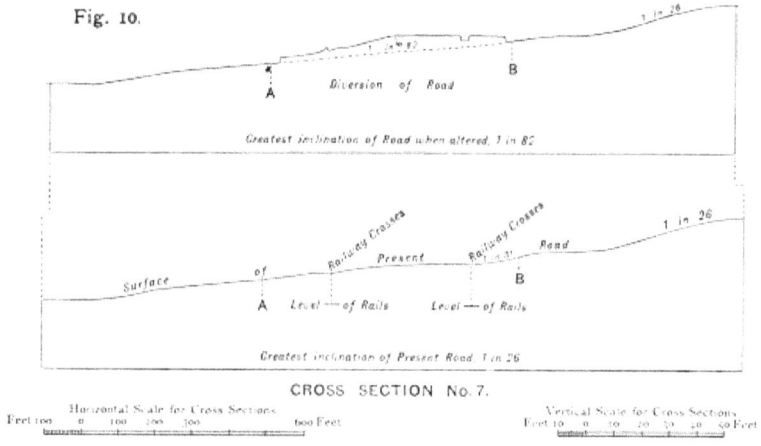

Home or starting signals next in advance of trading-points when lowered, to lock such points in either position, unless such locking will unduly interfere with the traffic.

A distant signal must not be capable of being lowered unless the home and starting signals in advance of it have been lowered.

8. Sidings to be so arranged that shunting operations upon them shall cause the least possible obstruction to the passenger lines. Safety-points to be provided upon goods and mineral lines and sidings, at their junctions with passenger lines, with the points closed against the passenger lines and interlocked with the signals.

9. When a junction is situated near to a passenger station, the platforms to be so arranged as to prevent, as far as possible, any necessity for standing trains on the junction.

10. The junctions of all single lines to be, as a rule, formed as double-line junctions.

11. The lines of railway leading to the passenger platforms to be arranged so that the engines shall always be in front of the passenger trains as they arrive at and depart from a station; and so that, in the case of double lines, or of passing places on single lines, each line shall have its own platform. At terminal stations a double line of railway must not end as a single line.

12. Platforms to be continuous, and not less than 6 feet wide for stations of small traffic, nor less than 12 feet wide for important stations. The descents at the ends of the platforms to be by ramps, and not by steps. Pillars for the support of roofs and other fixed works not to be less than 6 feet from the edges of the platforms. The height of the platforms above rail level to be 3 feet, save under exceptional circumstances, and in no case less than 2 feet 6 inches. The edges of the platforms to overhang not less than 12 inches. As little space as possible to be left between the edges of the platforms and those of the footboards on the carriages. Shelter to be provided on every platform, and conveniences where necessary. Names of stations to be shown on boards and on the platform lamps.

13. When stations are placed on, or near a viaduct, or bridge under the railway, a parapet or fence on each side to be provided of sufficient height to prevent passengers, who may by mistake leave the carriages when not at the platform, from falling from the viaduct or bridge in the dark.

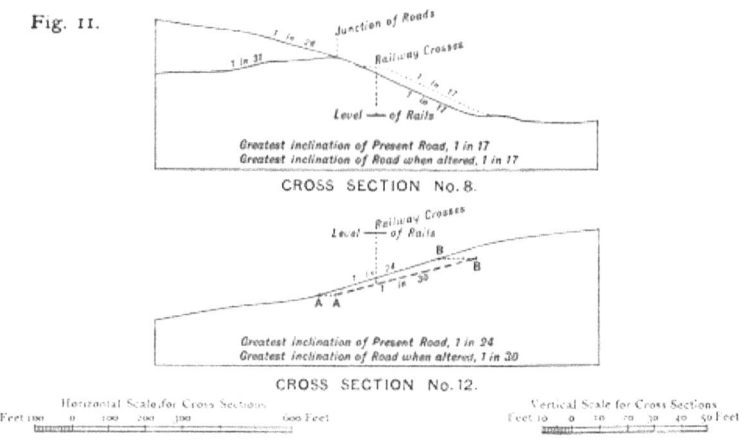

Fig. 11.

14. Footbridges or subways to be provided for passengers to cross the railway at all exchange and other important stations. Staircases or ramps leading to or from platforms to be at no point narrower than at the top, and the available width to be in no case contracted by any erection or fixed obstruction whatever below the top.

At all stations where crowding may be expected, the staircases or ramps to be of ample width, and barriers for regulating the entrance of the crowd at the top to be erected. If in such cases there are gates at the bottom, a speaking-tube or other means of communication between the top and bottom to be provided; and in all cases gates at the bottom of a staircase or ramp to open outwards. For closing the openings at the top, sliding bars or gates are considered best.

The steps of staircases to be never less than 11 inches in the tread, nor more than 7 inches in the rise, and midway landings to be provided where the height exceeds 10 feet.

Efficient handrails to be provided on both staircases and ramps, and in subways where ramps are used the inclination not to exceed 1 in 8.

15. A clock to be provided at every station, in some conspicuous position visible from the platforms.

16. No station to be constructed, and no siding to join a passenger line, on a steeper gradient than 1 in 260, except where it is unavoidable. When the line is double, and the gradient at a station or siding-junction is necessarily steeper than 1 in 260, and when danger

is to be apprehended from vehicles running back, a catch-siding with points weighted for the siding, or a throw-off switch, to be provided to intercept runaway vehicles at a distance outside the home-signal for the ascending line, greater than the length of the longest train running upon the line.

Under similar circumstances, when the line is single, provision for averting danger from runaway vehicles to be made—

(1) At a station in one of the following manners:—
   (*a*) A second line to be laid down, a second platform to be constructed, and a catch-siding or throw-off switch to be provided on the ascending line inside the loop-points.

   (*b*) A loop-line to be constructed lower down the incline than the station platform with a similarly placed catch-siding or throw-off switch.

(2) At a siding-junction in one of the following manners, except where it is possible to work the traffic with the engine at the lower end of a goods or mineral train, in which case an undertaking (see No. 35) to do so, given by the company, will be accepted as sufficient:—

   (*a*) A similar loop to be constructed as in the case of a station.

   (*b*) Means to be provided for placing the whole train on sidings clear of the main line before any shunting operations are commenced.

17. Engine-turntables of sufficient diameter to enable the longest engines and tenders in use on the line to be turned without being uncoupled to be erected at terminal stations and at junctions and other places at which the engines require to be turned, except in cases of short lines not exceeding 15 miles in length, where the stations are not at a greater distance than 3 miles apart, and the railway company gives an undertaking (see No. 35) to stop all trains at all stations. Care to be taken to keep all turntables at safe distances from the adjacent lines of rails, so that engines, waggons, or carriages, when being turned, may not foul other lines or endanger the traffic upon them.

18. Cast-iron must not be used for railway under-bridges, except in

the form of arched-ribbed girders, where the material is in compression.

In a cast-iron arched bridge, or in the cast-iron girders of an overbridge, the breaking weight of the girders not to be less than three times the permanent load due to the weight of the superstructure, added to six times the greatest moving load that can be brought upon it.

In a wrought-iron or steel bridge, the greatest load which can be brought upon it, added to the weight of the superstructure, not to produce a greater strain per square inch on any part of the material than five tons where wrought-iron is used, or six tons and a half where steel is used.

The engineer responsible for any steel structure to forward to the Board of Trade a certificate to the effect that the steel employed is either cast-steel, or steel made by some process of fusion, subsequently rolled or hammered, and of a quality possessing considerable toughness and ductility, together with a statement of all the tests to which it has been subjected.

19. In cases where bridges or viaducts are constructed wholly or partially of timber, a sufficient factor of safety, depending on the nature and quality of the timber, to be provided for.

N.B.—The heaviest engines, boiler trucks, or travelling cranes in use on railways afford a measure of the greatest moving loads to which a bridge can be subjected. The above rules apply equally to the main transverse girders and rail-bearers.

20. It is desirable that viaducts should, as far as possible, be wholly constructed of brick or stone, and in such cases they must have parapet walls on each side, not under 4 feet 6 inches in height above the rail level, and not less than 18 inches thick.

Where it is not practicable to construct the viaducts of brick or stone, and iron or steel girders are made use of, it is considered best that in important viaducts the permanent way should be laid between the main girders. In all cases substantial parapets, with a height of not less than 4 feet 6 inches above rail-level must be provided by an addition to the girders, unless the girders themselves are sufficiently high. On important viaducts where the superstructure is of iron, steel, or timber, substantial outside wheel-guards to be fixed above the level of, and as close to the outer rails as possible, but not so as to be liable to be struck by any part of an engine or train running on the rails.

In the construction of the abutments or piers which support the girders of high bridges and viaducts, cast-iron columns of small size must not be used.

In all large structures a wind-pressure of 56 lbs. per square foot to be assumed for the purpose of calculation, which will be based on the rules laid down in the report, dated 30th May, 1881, of the committee appointed by the Board of Trade to consider the question of wind-pressure on railway structures.

21. The upper surfaces of the wooden platforms of bridges and viaducts to be protected from fire.

22. All castings for use in railway structures to be, where practicable, cast in a similar position to that which they are intended to occupy when fixed.

23. The joints of rails to be secured by means of fish-plates, or by some other equally secure fastening. On main lines, and lines where heavy traffic may be worked at high speed, the chairs not to weigh less than 40 lbs.; but on branch lines, or lines on which the traffic is light, chairs weighing not less than 30 lbs. may be used.

24. When chairs are used to support the rails they must be secured to the sleepers, at least partially, by iron spikes or bolts. With flat-bottomed rails, when there are no chairs, or with bridge rails, the fastenings at the joints, and at some intermediate places, to consist of fang or other through-bolts; and such rails, on curves with radii of 15 chains or less, to be tied to gauge by iron or steel ties at suitable intervals.

25. In any curve where the radius is 10 chains or less, a check-rail to be provided.

26. Diamond-crossings, as a rule, not to be flatter than 1 in 8.

27. No standing work (other than a passenger platform) to be nearer to the side of the widest carriage in use on the line than 2 feet 4 inches, at any point between the level of 2 feet 6 inches above the rails, and the level of the upper parts of the highest carriage doors. This applies to all arches, abutments, piers, supports, girders, tunnels, bridges, roofs, walls, posts, tanks, signals, fences, and other works, and to all projections at the side of a railway constructed to any gauge.

28. The intervals between adjacent lines of rails, where there are two lines only, or between lines of rails and sidings, not to be less than 6 feet. Where additional running lines of rails are alongside the main lines, an interval of not less than 9 feet 6 inches to be provided, if possible, between such additional lines and the main lines.

29. At all level crossings of public roads, the gates to be so constructed that they may be closed either across the railway, or across the road at each side of the crossing, and a lodge, or, in the case of a station, a gatekeeper's box, to be provided, unless the gates are worked from a signal cabin. The gates must not be capable of being opened at the same time for the road and the railway, and must be so hung as not to admit of being opened outwards towards the road. Stops to be provided to keep the gates in position across the road or railway. Wooden gates are considered preferable to iron gates, and single gates on each side to double gates. Red discs, or targets, must be fixed on the gates, with lamps for night use, and semaphore signals in one or both directions interlocked with the gates, may be required. At all level crossings of public roads or footpaths, a footbridge or a subway may be required.

At occupation and field crossings, the gates must be kept hung so as to open outwards from the line.

30. Sidings connected with the main lines near a public road level crossing to be so placed that shunting may be carried on with as little interference as possible with the level crossing; and, as a rule, the points of the sidings to be not less than 100 yards from the crossing.

31. At public road level crossings in or near populous places, the lower portions of the gates to be either close barred, or covered with wire netting.

32. Mile posts, half-mile, and quarter-mile posts, and gradient-boards to be provided along the line.

33. Tunnels and long viaducts to be in all cases constructed with refuges for the safety of platelayers. On under-bridges without parapets, handrails to be provided. Viaducts of steel, iron, or timber to be provided with manholes or other facilities for inspection.

34. Continuous brakes (in accordance with the Regulation of Railways Act of 1889), complying with the following requirements, to be provided on all trains carrying passengers, viz.—

(1) The brake must be instantaneous in action, and capable of being applied by the engine-driver and guards.

(2) The brake must be self-applying in the event of any failure in the continuity of its action.

(3) The brake must be capable of being applied to every vehicle of the train, whether carrying passengers or not.

(4) The brake must be in regular use in daily working.

(5) The materials of the brake must be of a durable character, and easily maintained and kept in order.

35. Any undertaking furnished by a railway company to be under the seal, and signed by the chairman and secretary of the company.

Recommendations as to the Working of Railways.—1. There should be a brake vehicle, with a guard in it, at or near the tail of every passenger train; this vehicle should be provided with a raised roof and extended sides, glazed to the front and back, and it should be the duty of the guard to keep a constant look-out from it along his train.

2. All passenger carriages should be provided with continuous footboards, extending the whole length of each carriage and as far as the outer ends of the buffer castings. As passenger carriages pass from one company's line to another's, it is essential for the public safety that, although the widths of the carriages on the different lines may differ from each other, the widths across the carriages from the

outside of the continuous footboard on one side, to the outside of the continuous footboard on the opposite side, should be identical for the carriages of all railway companies, so that the lines of rails may be laid at the proper distance from the edges of the passenger platforms.

3. There should be efficient means of communication between the guard, or guards, of every passenger train and the engine-driver, and between the passengers and the servants of the company in charge of the train.

4. The tyres of all wheels should be so secured as to prevent them from flying open when they are fractured.

5. The engines employed with passenger trains should be of a steady description, with not less than six wheels, with the centre of gravity in front of the driving-wheels, and with the motions balanced. They should, as a rule, be run chimney in front.

6. Records should be carefully kept of the work performed by the wearing parts of the rolling stock, to afford practical information in regard to them, and to prevent them from being retained in use longer than is desirable.

7. In addition to the block-telegraph instruments, it is desirable that there should be speaking-instruments, or telephones, for communication between signalmen, and books for recording the running of the trains.

8. When drovers or other persons are permitted to travel with goods or cattle trains, suitable vehicles should be provided for their accommodation.

9. It is considered that, in fixed signals, the front lights should show—
    Green, for all right;
    Red, for danger;
and that back lights, visible only when the signals are at danger, should show white.

10. Refuge sidings should be provided at all main-line stations where slow trains are liable to be shunted for fast trains to pass them. If at such stations it is impossible to provide refuge sidings, and slow trains have to be shunted from one main line to the other to allow of

fast trains passing them, some simple arrangements should be supplied in the signal cabins to help to remind the signalman of the shunted train.

11. Efficient means should be adopted to prevent the accidental opening of the doors of passenger trains.

To carry out the undertaking, the engineer has to prepare working plans and sections to a somewhat larger scale than that adopted for the Government or Parliamentary plans, and on which must be marked the exact positions of the commencement of the curves, straight lines, and gradients. The sites of all the over and under bridges must be shown, and their angles of crossing noted. All road, river, or stream diversions must be indicated, so that the work in connection with them may be laid out on the ground. All culverts and drains must be marked, and their size, depth, and direction described. Public road level-crossings, and farm or occupation-road crossings, must be shown in their proper positions.

The face-lines of the ends of all tunnels should be marked on the working plan and section, and the position of any shafts, which may be intended either for use in carrying on the work or for future ventilation.

A considerable amount of investigation and negotiation will have to be entered into before the locating of the above works can be finally decided. The desire to meet the wishes and convenience of all parties interested must of necessity be controlled by the physical circumstances of each case; very little alteration can be made in the level of the rails, although some variation may be made in their position.

When fixing the depths of culverts and drains, attention must be paid to any probable improvement in the drainage of the district, which might at some future time necessitate the deepening of such of the main culverts where the inverts had been laid too high.

Unless all these details are determined, and shown on the working-plans before the works are commenced, there is the risk that embankments may have to be opened out to admit of bridges and culverts, and cuttings changed to permit of road diversions.

The entire centre-line of railway must be carefully staked out by driving strong wooden pegs into the ground at the end of every chain length, and along the course of these pegs the longitudinal section must be taken. Three pegs, one on each side of the centre peg, are generally placed at the commencement and termination of the curves. When the longitudinal section has been plotted to scale, and the course of the gradients and level portions worked out and drawn on, then the heights of the ground level and formation level can be marked at each chain, and from them the depths of the cutting and the heights of the embankments can be ascertained and marked at each chain. In addition to the longitudinal section, it will be necessary to take a large number of transverse or cross sections at those pegs, or intermediate points, where the ground is at all side-lying or irregular. These cross-sections are necessary to determine the side-widths, or distances to outer edge of slopes in cuttings or embankments, and also to calculate the actual quantity of earthwork to be executed. For convenience in taking out the quantities, these cross-sections are generally plotted to a natural scale, that is to say, to the same scale horizontal as vertical, as shown in the example of cross-sections, Figs. 15 to 24. It is also necessary to obtain information, by boring or otherwise, as to the material of which the cuttings are composed, whether clay, gravel, or rock.

In laying out lines through fairly level plains and populous districts, the absence of great natural obstacles will allow the engineer to carefully consider how far it may be prudent to diverge to the right or to the left, to accommodate towns and places which would be excluded by a more direct through route. There will be ample range for selection, and it will be rather the question of policy than compulsion which will guide him in the route to be taken.

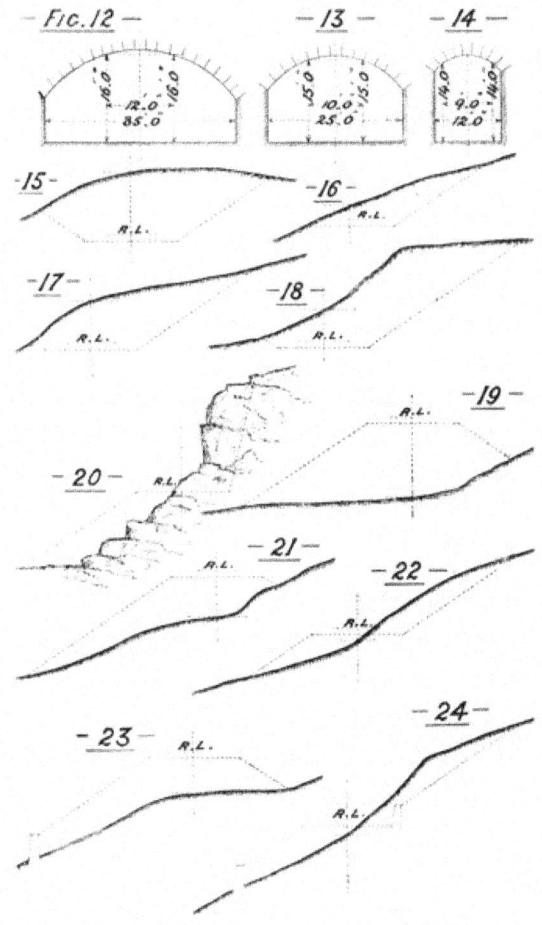

When, however, the locating passes from the lower ground, away up amongst the hills and mountain ranges, it becomes an intricate study whether it will be possible to lay out any line at all which may possess gradients and curves practicable for railway working. The question of property, population, or convenience of access, is here no longer the controlling influence, but in its stead there are the far more formidable natural difficulties to be overcome in working out a trackway to the inevitable summit level. The chief endeavour will be to gain length, and so reduce as much as possible the steepness of the gradients which at the best must necessarily be severe. In some of the earlier mountain lines constructed abroad the system of *zigzags* was introduced, as shown in Fig. 25. These *zigzags* were laid out on ruling gradients, one above the other, on the sides of the mountain slopes with pieces of level at the apices, **A**, **B**, and **C**, on which the engine could be changed from one end of the train to the other. Although feasible in principle, the system entailed considerable loss of time in train-working, and was not unattended with risk.

The more modern and simple method of working out the same idea is to connect the main zigzag lines by curves or *spirals*, thus rendering the route continuous and unbroken. By this arrangement the heavy work and delay in starting or stopping the train at the apices, **A**, **B**, and **C**, as shown on Fig. 25, is avoided, and the train can proceed continuously on its circuitous journey. Fig. 26 shows an instance of the zigzags and spirals, as carried out on an important railway abroad. To have made a direct line from **D** to **E**, the most difficult part of the route, would have involved a gradient of 1 in 11; but by constructing the spiral course, as shown, the length was more than trebled, and the gradient reduced to 1 in 35.

Fig. 27 is another example of spiral zigzags in which advantage was taken to cut a short tunnel through a high narrow neck of rock at **G**, and then by skirting round the hill the line was taken over the top of the tunnel and along the side of the mountain to the summit tunnel at **H**. By this means the line from **F** to **H** was laid out to an average gradient of 1 in 42.

Fig. 28 shows the Cumbres inclines on the Mexican Railway. The route had to be located through one of the rugged passes of the great Chain of the Andes, whose mountain-sides rise most abruptly from the lower plains, to the great upper-land plateau, some eight thousand feet above sea-level. The ground to be traversed was so steep and difficult that, even with the best available detours and greatest length that could be obtained, the result was an average continuous gradient of 1 in 25 for 12 miles.

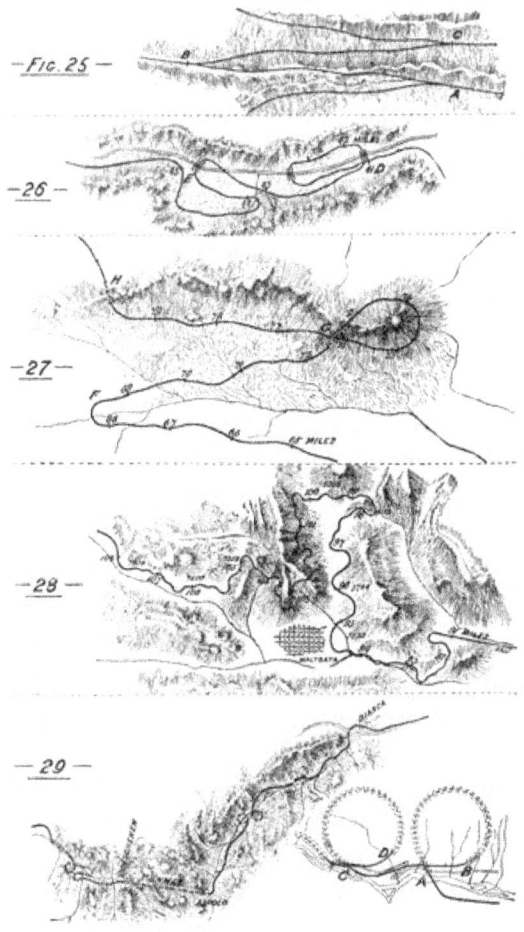

Fig. 29 is a plan of part of the St. Gothard Railway, showing the principal tunnel 9¼ miles long, and some of the adjoining spiral tunnels. The long tunnel through the great Alpine barrier was the only means of forming a railway connection between the two points at Airolo and Goeschenen. Constructed in a straight line, with easy gradients, falling towards the entrances, efficiency of drainage has been secured, and excessive strain on motive-power avoided. The approaching valleys on each side were in some places too irregular and broken to admit of zigzag loops, and the spiral tunnels were adopted instead. The enlarged plan of two of the spiral tunnels will explain the method of working. An ascending train enters the first tunnel at **A**, and after passing round almost an entire circle, on a rising gradient, emerges at a much higher level at the point **B**. Proceeding onward, the train enters the second tunnel at **C**, and after passing round a similar circle, on a rising gradient, comes out at a still higher point, **D**, and continues its course up the valley.

The last five sketches illustrate some of the methods which have been adopted when constructing railways through some of the most difficult mountain ranges. They show what has been done, and may serve as guides in working out the location of a line in some hitherto unexplored region.

Gauge.—The gauge of a railway, or its width from inside to inside of rails, affects both its cost and efficiency. If the gauge be exceptionally wide, then the expenditure on works and rolling-stock will be proportionately heavy; and although theoretically the extra wide gauge may possess greater capabilities for accommodation and high-speed travelling, we may find in practice that the necessary requirements may be provided on a much more moderate gauge. On the other hand, if the gauge be exceptionally narrow, there will be diminished convenience both for passengers and merchandise, and a corresponding limit to the speed in transit.

In isolated districts, where passenger traffic is of secondary importance, and where the principal merchandise will be heavy without being bulky, such as mineral ores, slates, etc., a comparative narrow gauge may possibly suit the purpose. For main trunk lines, however, where a large, heavy, and fast passenger traffic will have to be worked, and where goods of all kinds, many of them bulky without being heavy, will have to be carried, an ample gauge must be selected to ensure convenience and safety. A liberal gauge permits the use of commodious rolling-stock without any great amount of lateral overhanging weight outside the wheels; whereas with a narrow gauge there is the tendency—if not the necessity—to use vehicles which have too great a lateral overhang for proper stability, except at very moderate speeds.

The following list shows the gauges adopted in various countries:—

ft. ins.

England, Scotland, and Wales	4	8½
Ireland	5	3
United States	4	8½, with some lines 5 ft., 5 ft. 6 ins., and 6 ft.
Canada	4	8½ and 5 ft. 6 ins.
France	4	8½
Belgium	4	8½
Holland	4	8½
Germany	4	8½
Austria	4	8½
Switzerland	4	8½
Italy	4	8½
Turkey	4	8½
Hungary	4	8½
Denmark	4	8½
Norway	4	8½ and 3 ft. 6 ins.
Sweden	4	8½
Mexico	4	8½ and 3 ft.
Egypt	4	8½ and 3 ft. 6 ins.
Peru	4	8½
Nova Scotia	4	8½ and 5 ft. 6 ins.
New South Wales	4	8½
Brazil	4	8½, 5 ft. 3 ins., and 5 ft. 6 ins.
Uruguay Republic	4	8½
Russia	5	0
South Australia	5	3
New Zealand	3	6

British India	5	6	and 1 metre.
Ceylon	5	6	
Spain	5	6	
Portugal	5	6	
Chili	5	6	
Argentine Republic	5	6	
Cape Colonies	3	6	
Japan	3	6	

After many years' experience of actual working, the broad, 7 feet, gauge of the Great Western Railway has been abandoned for the 4 feet 8½ inch gauge. Doubtless this decision was the result of most careful deliberation, and was made upon convincing proof that the 4 feet 8½ inch gauge could fulfil all the advantages claimed for the wider gauge, whilst at the same time it possessed the merit of less cost of construction and working, and greater facilities for the exchange of traffic with other lines having the standard gauge. The facility of exchange, or through working of rolling-stock, is a leading element of successful railway working, and it is difficult to estimate what would be the amount of loss and delay if we had any great extent of break of gauge on the main trunk lines of our own country.

Although some countries have selected gauges of 5 feet and 5 feet 6 inches, it is interesting to note that the largest number have adopted the English standard gauge of 4 feet 8½ inches, and that the miles of line laid to this gauge far outnumber all the others. The fact that our own home lines, the principal Continental lines, and nearly all that vast network of railways in the United States of America, have been laid to the 4 feet 8½ inch gauge, testifies to the general opinion of its utility and efficiency; and we know that included in that list are the railways which carry the largest, heaviest, and fastest train service in the world.

It would be interesting to trace back, and, if possible, ascertain from whence the exact gauge of 4 feet 8½ inches was derived. No doubt, in the early days of the pioneer iron highways in England, the railways were made the same gauge as the tramroads which they superseded. But why was 4 feet 8½ inches the gauge of the tramroads? We may reasonably infer that the first four-wheeled waggons used on the early tramroads were in reality the same waggons which had been previously used on the common roads for the conveyance of coal and minerals to the ports for shipment, and that the waggons were merely transferred from the roughly paved or macadamised roads to the tramroads. Flanged wheels were then unknown, and the introduction of the tram-plates was at first simply designed to lessen the resistance to haulage. The gauge, or width between the wheels, of these waggons may have been the outcome of long experience as to the most suitable width for convenience of load, stability during transit, or for space occupied on the highway. The width may have been handed down from generation to generation, going back to the time when wheeled vehicles were first built in the country. Perhaps in the beginning the first vehicles may have been imported from Italy, or Greece—countries which in the earlier ages were the most advanced in matters of luxury and convenience.

When in Pompeii, a few years ago, the writer measured the spaces between a large number of the *wheel-ruts* which are worn deep into the paving-stones in many of the principal streets of that wonderful unearthed city. These paving-stones, very irregular in shape, and many of them 2 feet 6 inches long by 1 foot 6 inches wide, are carefully fitted together, and form a compact massive pavement from curbstone to curbstone. The wheel-tracks, which are in many places worn into the stones to the depth of an inch or an inch and a half, are always distinct, and there is no difficulty in defining the corresponding track.

The result of a large number of measurements gave an average width of about 4 feet 11 inches from centre to centre of the wheel-tracks, a curious coincidence with the gauge of our own road vehicles at the beginning of the railway era. Whether our selection of the railway gauge of 4 feet 8½ inches has been the result of study, imitation, or caprice, we certainly have the silent testimony of these old deep-worn stones to prove that two thousand years ago the chariots of Pompeii were of very similar gauge to our own of modern times.

Narrow-gauge railways, of gauges varying from 1 foot 10½ inches on the Festiniog Railway, to 3 feet, 3 feet 3 inches (metre), and 3 feet 6 inches, have been made in several places both at home and abroad. Generally speaking, they have been constructed as subsidiary or auxiliary lines in thinly populated districts, with a view to afford some railway accommodation where it was considered that lines of the standard gauge would not pay. In some instances abroad long lines of narrow gauge—3 feet and 3 feet 6 inches—have been constructed as main trunk lines in newly opened out districts. Some of these have since been altered to a wider gauge as the traffic developed, and experience proved that the narrow width of the vehicles was unsuitable for quick transit, or convenience in the accommodation of passengers and goods.

The object in making a line to a narrow gauge is doubtless to save cost in the original construction; but when a scheme for an altered gauge is put forward, it will be well to consider what amount of advantage or saving would be effected by deviating from the standard gauge.

If there be almost a certainty that such proposed line will always remain isolated from all other existing railways of the standard gauge, then perhaps the selection of gauge may be one of minor importance, and there remains but the question whether the description of traffic, and the weights to be carried, can be worked to any greater advantage, or more economically, by deviating from the standard gauge.

If, however, there be a fair probability that such proposed line may at some future time become part of an already established railway system, it would appear to be more prudent to make the line to the standard gauge, and effect economies by introducing steeper gradients, sharper curves, and lighter permanent way, and keep down working expenses by using lighter locomotives, worked at slower speeds.

High speeds are not expected on narrow gauge railways, and no complaints are made about passenger trains whose highest running speed does not exceed 20 miles per hour. By conceding the same indulgence to light railways made to the standard gauge, great economies might be introduced both in their construction and working. The similarity of gauge would admit the transit of the carriages and waggons of other standard gauge lines, and so avoid all cost and delay in transshipment. The heavy engines could be kept for the main-line working, and light engines for slow speeds would serve for the light standard-gauge lines. As traffic developed, and the train service required heavier and faster trains, the light rails could be removed, and replaced by those of heavier section to correspond to the main line. The similarity of gauge would permit uninterrupted transit of all vehicles to a common centre for repairs, whereas the narrow gauge carriages and waggons, being limited to running only on their own district, must have separate workshops for their repair.

When considering the cost of construction and working of a narrow-gauge railway as compared with one of the standard gauge, there are certain items which are common to both, and in which the narrow gauge could not be expected to obtain any advantage over the standard gauge.

There would not be any saving in getting up the scheme in the first instance;
Nor in the Parliamentary expenses;
Nor in the engineering or carrying out of the works;
Nor in the station accommodation, waiting-rooms, and offices;
Nor in the signals and interlocking arrangements;
Nor in the telegraph;
Nor in the working staff and train men;
Nor in the maintenance of the permanent way, as the same number of men would be required for the inspection and packing of the road, perhaps more.

Little or no saving could be expected in the bridges under the railway, as these must be made to the prescribed widths and heights, irrespective of the gauge of the railways.

Little, if any, saving could be made in river or stream bridges, as the same amount of waterway would have to be provided in each case.

The same remark applies to culverts and drains.

There would, on the other hand, be a small saving in the quantity of land to be acquired to the extent of a narrow strip or zone, represented by the difference in width between the narrow and standard gauges.

There would also be the same small proportionate saving in the embankments and cuttings to the extent of the difference in gauge.

Also a saving in the overline bridges and road approaches in consequence of less width and height of the opening through those bridges.

And a saving in the rails, sleepers, and ballast of the permanent way, to the extent consistent with efficiency. That some saving may be effected in these is undoubted, but it is necessary to exercise caution, and not rush to the opposite extreme by making the parts too light. A rail should be made not only strong enough to carry well the engines that have to pass over it, but it should also be heavy enough to stand the wear of several years. Narrow-gauge engines must be heavy in conformity with the loads they have to haul. The same amount of power must be exerted to haul a hundred tons on a given gradient, whether the gauge be narrow or broad. In some cases of narrow-gauge railways the original rails, which weighed only 45 lbs. per yard, have since been replaced with others weighing 60 and 65 lbs. per yard. The light 45 lb. rails were evidently not found to be sufficiently heavy to keep the road to proper line and level. The result of our everyday practice seems to prove that there is not only an advantage, but an economy, in adopting rails of a heavy section, and experience would therefore indicate that even for a narrow-gauge railway it may not be expedient to adopt rails weighing less than 65 lbs. per yard.

Gradients.—There are very few localities where the rails on any line of railway can be laid perfectly level or horizontal for more than comparatively short distances. By far the greater portion have to be laid on inclined planes of varying rates of inclination to suit the general formation of the district traversed, and the circumstances of the line to be constructed.

The degree, or rate of inclination, of these inclined planes, or gradients, may be expressed in various ways. A very general method is to state the number of feet, metres, etc., which can be measured along the gradient before an increased rise or fall of one foot or metre, etc., is obtained. Thus a gradient of 1 in 200 signifies a rise or fall of 1 foot in 200 feet, or 1 metre in 200 metres.

Sometimes the rate of inclination is expressed by stating the number of feet of rise or fall in a mile. In this way a gradient would be described as falling at the rate of 30 feet in a mile, rising at the rate of 20 feet in a mile, etc. Twenty feet to a mile is equal to 1 in 264.

Another method is to give the percentage of rise or fall. In this way the inclination would be expressed as a 1 per cent. gradient, 2 per cent. gradient, ½ per cent. gradient, etc., which for comparison would signify 1 in 100, 1 in 50, and 1 in 200 respectively.

The gradients of a railway most materially influence its facility and cost of working, and every effort should be used to make them as easy as possible consistent with the prospect of the line.

Steep gradients signify heavy locomotives, increased cost of motive-power, reduced speed, and light loads.

The following tabulated memoranda show the approximate loads, exclusive of engine and tender, which can be hauled on the level and on certain inclines at various speeds by engines of the quoted capacities and steam admissions. A medium-sized, ordinary type of passenger and goods engine has been selected for each of the examples. The working of the passenger engine and train is assumed to be under favourable circumstances, with fine weather, fairly straight line, first-class permanent way, modern rolling-stock with oil axle-boxes and perfect lubrication, and all the conditions most suitable to ensure the least resistance to the moving load. For the goods engine and train a greater resistance per ton of load is assumed, as the goods trucks are never so perfect or easy in the running as the passenger carriages. A certain amount of side wind is taken into consideration, and also an allowance for moderately sharp curves, the object being to indicate what may be looked upon as fair, average, workable loads.

The loads for engines of larger or smaller dimensions, or higher or lower pressures, may be obtained by working out the proportion between the tractive force put down in any of the columns of the tabulated memoranda and the ascertained tractive force of any other engine under the same conditions of cut-off and speed.

	PASSENGER ENGINE.			
	Six wheels, driving and trailing wheels coupled, 6 ft. 6 ins. diameter. Cylinders, 17 ft. × 24 ft. Locked-down pressure on safety-valves, 140 lbs. per square inch. Assumed pressure at cylinders, 120 lbs. per square inch. Weight of engine 39 tons. " tender <u>24</u> tons. 63 tons.			
Assumed cut-off	¼	⅓	½	¾
" mean effective pressure, lbs.	45	56	76	100
" tractive force, lbs.	4000	4979	6758	8892
Speed in miles per hour	60	40	30	15
	Tons.	Tons.	Tons.	Tons.
Level	97	230	447	892
1 in 1000	84	196	373	707
" 800	81	188	358	671
" 600	76	177	335	618
" 400	68	157	296	533
" 300	60	141	263	467
" 250	55	129	241	424
" 200	47	114	213	372
" 150	37	93	177	304
" 100	21	63	126	217
" 90	—	56	114	197
" 80	—	48	101	175
" 75	—	43	94	164
" 70	—	39	86	152
" 60	—	28	70	128
" 50	—	—	53	101
" 40	—	—	—	73
" 25	—	—	—	27

	GOODS ENGINE.
	Six wheels, all coupled, 4 ft. 6 ins. diameter. Cylinders, 17 ft. × 24 ft. Locked-down pressure on safety-valves, 140 lbs. per square inch. Assumed pressure at cylinders, 120 lbs. per square inch.
	Weight of engine 34 tons. " tender <u>24</u> tons. 58 tons.

Assumed cut-off	¼	⅓	½	¾
" mean effective pressure, lbs.	45	56	76	100
" tractive force, lbs.	5780	7192	9760	12844
Speed in miles per hour	40	30	20	15
	Tons.	Tons.	Tons.	Tons.
Level	213	358	623	907
1 in 1000	187	310	532	768
" 800	181	299	512	739
" 600	172	285	482	695
" 400	157	257	432	621
" 300	143	233	390	560
" 250	133	216	361	519
" 200	120	195	324	467
" 150	101	165	276	397
" 100	74	123	208	302
" 90	—	113	191	279
" 80	—	101	172	253
" 75	—	95	163	240
" 70	—	88	153	226
" 60	—	74	131	196
" 50	—	—	107	163
" 40	—	—	—	127
" 25	—	—	—	67

NOTE.—The column loads in tons are exclusive of the weight of engine and tender.

From the above memoranda it will be seen how greatly the gradients affect the loads. For an important main trunk line, with a heavy and frequent train-service of passengers and goods, the introduction of steep gradients would not only reduce the speed of the train-working, but would probably involve the necessity of assistant engines over those parts of the line; and it may be prudent, where possible, to incur heavier earthworks, or considerable detours, or tunnels, to obtain more favourable gradients. For such a line the additional cost, and the extra distance caused by a detour of a mile or more, will be of far less importance than the interruption in the train service arising from a serious reduction in speed or taking on assistant engines. On many railways abroad there are very interesting examples of long detours of several miles, carefully studied out to obtain greater length and easier gradients, resulting in the construction of lines over which the traffic can be worked without necessitating auxiliary engine-power. On the other hand, there are situations where steep gradients cannot be avoided, where certain altitudes must be reached, and where there is no alternative but to face the inevitable.

On secondary lines, and short branch lines, where the traffic is not expected to be heavy, and where speed is not so important, it may be policy to economize outlay and introduce steeper gradients than on the main line.

Half a mile of a rather steep gradient is not felt so much when it is situate midway between two stations, because the attained speed of the train assists the engine over the short distance to the summit; but when it occurs as a rising gradient out of a station, it forms a great check to the working, particularly in bad or wet weather, when there is the risk of the engine slipping, and the entire train sliding back into the station.

Long steep gradients not only necessitate increased motive-power for the ascending trains, but also require increased brake-power, and precautionary measures for the descending trains. Where passenger trains are fitted with continuous brakes, the risk of losing control is minimized; but with goods trains composed of waggons, having only the ordinary independent side-lever brake, it will be found absolutely necessary in many cases to have additional heavy brake-vans for descending the inclines, and these special vans, unfortunately, will form so much extra non-paying weight to be hauled up on the ascending trains. Of course, it is quite possible—and, indeed, in many places it is customary—to pin down some of the side-lever brakes before commencing the descent, but once pinned down the brakes cannot be eased or taken off until the entire train is brought to a stand.

Every goods waggon should be fitted with a brake, and it would be of immense value if that brake could in all cases be applied and controlled when the train is in motion.

The American type of long goods waggon, with a four-wheel bogie-truck at each end, is fitted with a brake very similar to those adopted on the ordinary horse tram-cars. On the top of the waggon a horizontal iron hand-wheel, about 18 inches in diameter, is fixed on to a strong vertical iron rod, which works in brackets, and extends down below the underside of waggon framing. One end of a short length of chain is secured to the foot of the vertical rod, and the other end is connected by light iron rods to the series of levers which pull on the brake-blocks. By rotating the horizontal hand-wheel the chain is coiled round the lower end of the vertical rod, the brake-levers are pulled over, and brake-pressure applied to the wheels of the waggon. The brakesman is supplied with a convenient seat and footboard, and on the floor-level of the latter there is a pawl and ratchet attached to the vertical rod, which permits the brakes to be applied to the extent required. The pawl retains the brakes in position until the brakesman with his foot pushes the pawl out of the notch of the rachet and releases the brake gearing, which is at once pulled off quite clear by strong bow-strings attached to the framework of the bogies.

This type of hand-brake is, perhaps, the simplest that can be made. The brakesman has merely to put it on, the pawl and ratchet keep it on, and the bow springs take it off when no longer required. Each one of these long, loaded goods waggons becomes a very serviceable brake-van, and for ascending and descending steep inclines all that is necessary is to take on a few additional brakesmen to manage the brakes of as many suitable waggons. These incline brakesmen, after going down, can return to the summit by the next ascending train, their small weight being a mere nothing as compared with that of special or extra brake-vans.

On some European lines it is the custom to sprag some of the goods waggon wheels when going down exceptionally steep inclines, as well as applying the brakes on the ordinary and extra brake-vans. The sprag is a piece of wood, circular in section, about 2 feet 6 inches long, and 5 to 6 inches thick in the middle, tapering off to about 2 inches thick at the ends. When the waggon-wheel is just beginning to move, the sprag is inserted between the spokes, and being caught against the waggon framework, the wheel is held fast, and being unable to revolve, remains fixed, and acts like a skid upon the rails. The skidding of the wheels upon the rails wears flat places on the wheel tyres, and it is needless to mention that the practice is only resorted to in very extreme cases. Although a very primitive means for checking the speed of a descending train, or for maintaining vehicles stationary on an incline, there have been many instances where lives have been saved and accidents prevented by the prompt use of a few sprags. Solid or close wheels cannot be spragged, only wheels which have spokes or openings, and for this reason alone it would be very desirable that in every passenger and goods train there should be some spoke or open wheels which could be spragged as a last resource, in the event of a sudden emergency of brakes failing or train becoming divided on an incline.

On ascending gradients there is always the risk of a coupling breaking, and the train becoming divided. If the detached portion left behind be provided with ample brake-power, hand-brakes, or otherwise, no harm may take place beyond a little delay; but if the brake-power be insufficient or defective, and if all the wheels are solid wheels incapable of admitting a few timely sprags, then the vehicles cannot be held, but must slide back, and running unchecked would soon attain such a velocity as would cause them either to leave the rails or dash into another train standing at the last station. Many lamentable accidents have taken place arising from portions of trains breaking away and running back, and the sad experience of those casualties should call forth every effort to avert a recurrence in the future. It may not always be possible to detect a hidden flaw in a coupling, or a defect in the brake-gearing until the actual failure occurs; but it is quite possible to guard against disastrous results from such failure, by providing means to hold the vehicles, and prevent them sliding back.

For some years the writer had the entire charge of an important railway abroad on which the gradients were very exceptional, and where it was absolutely necessary that he should organize the most complete precautions to prevent the possibility of trains, or portions of trains, running back down inclines. Starting from sea-level, the line, which was laid to the 4 feet 8½ inch gauge, rose to a summit of over 8000 feet, and on the mountain division there were many long gradients of 1 in 40, 1 in 33, and in one place a continuous gradient of 1 in 25 for 12 miles. The specially powerful engines reserved for these heavy inclines were each supplied with an ordinary hand-brake, a steam-brake, and a Westinghouse continuous brake. The passenger carriages, which were of considerable length, and carried on a four-wheeled bogie-truck at each end, were all fitted up with the Westinghouse brake, and in addition each carriage had its own hand-wheel brake with the pawl and ratchet gearing. All the goods waggons, which were of the American type, were fitted with hand-wheel brakes similar to those on the carriages. Special gangs of trained brakesmen took charge of the trains on these inclines, a brakesman to every carriage or waggon, and were always in readiness in case of the breakage of a coupling, or the failure in the Westinghouse brake or brakes on engine. The immunity from accidents justified the combined precautions adopted, and proved the possibility of working such severe gradients with perfect safety.

The long-continued application of the brakes on heavy inclines naturally leads to the question as to the description of wheel to be adopted for the work. Not only are the wheels subjected to very severe torsional strains, but the temperature at the circumference is raised very high in consequence of the friction. Perhaps, theoretically, the safest wheel would be one made out of a solid piece of metal, similar to the chilled cast-iron wheels of the United States, or the steel disc wheels used on some lines in Europe, in either of which holes can be left for sprags. Wheels of this description can withstand very heavy wear and tear, they are not affected by increased temperature, and they certainly have the minimum of parts to work loose. Of the built-up wheels, the strong forged-iron-spoke wheel with steel tyres shows excellent results, and always gives due warning of loosening by indications at the tyre rivets. The suddenness with which the solid wooden centre wheels sometimes break up and fall to pieces does not commend them for a service where there must be a long-sustained application of the brakes. The increased temperature which expands the tyre, contracts the wood, and must loosen and weaken the entire wheel.

On all steep gradients the road-bed should be of the most substantial character, and the permanent way of a strong description, and maintained in perfect order, as the engines for working the traffic must necessarily be of a heavy type. The rails will be severely tested by the pounding and slipping of the engines on the ascending journey, and by the action of the brakes on the descending journey.

In the early days of the railway system, rope-haulage was adopted on some of the main lines for working the trains on steep inclines near the principal terminal stations. A powerful stationary engine, located at the highest point, was employed to work an endless rope which passed round large drums at the top and bottom of the incline, and was supported on sheaves or pulleys fixed between the rails. The connection between the carriages and endless rope was effected by means of a short piece of rope called the *messenger*, which was coiled round the main rope in such a manner as to be readily detached when the train reached the summit. There are many persons who will remember the time when the passenger trains were hauled by an endless rope up the 1 in 66 incline from Euston to Camden Town, a distance of about a mile and a half, and up the 1 in 48 incline from Lime Street, Liverpool, to Edge Hill, a distance of about a mile and a quarter, and several others. The rapid strides made in locomotive construction, and the increased pressure used in the boilers, enabled much more powerful engines to be built, until one by one the rope-haulage machinery has disappeared from nearly all the inclines where for years it had been considered indispensable. Rope-haulage on inclines is now very rarely met with, except at collieries and ironworks, where occasionally the rope may be seen so arranged that the loaded waggons descending pull up the empty waggons on the opposite or parallel line.

Curves.—The degree of curvature of a railway curve is generally expressed by giving the radius in feet, chains, metres, or other national standard measure.

When laying out a line of railway, the natural features of the country will necessitate the introduction of curves, and the question for consideration will be whether they are to be made of small or large radius. In some cases sharp curves are inevitable, except by incurring enormous works which would not appear to offer any corresponding prospective recompense. In others the curves may be made of easy radius, at a comparative moderate extra outlay, if the character of the line and description of traffic to be accommodated will warrant the expenditure. For main through lines, with heavy, high-speed traffic, it is advisable to have the curves of large radius, so as to avoid the necessity of reducing speed when passing round them. Although a high-class fast train may be allowed to run round an 80 chain (5280 feet) curve at almost unrestricted speed, safety demands that there should be a reduction of speed on curves of 40 or 30 chains radius, and a very much greater reduction for curves of 20 chains radius and under. A sharp curve will in some places form a greater check to fast trains than a length of moderately steep gradient on a straight line. In the former the trains running in either direction must slow down for some distance before reaching the curve, round which they should pass at greatly reduced speed, and then some distance must be run before they can attain their full speed again. On the other hand, with a rising gradient, on a fairly straight line, the acquired momentum of the train will materially assist in ascending the incline, and although the speed may be slackened as the train advances, there may not be any very great diminution in the running before the gradient is passed, and average level line reached again. A reduced rate of running must be maintained round curves of small radius, for, however substantial the works and permanent way, and however well devised and constructed the rolling-stock, there is an element of danger ever present when passing round sharp curves at anything more than moderate speed. In the great rush for fast through trains this point is very apt to be overlooked, and too little time allowed for the running. Even with the fastest trains on any line there are some portions of the route which must be traversed with greater caution and less speed than others, either on account of sharp curves or of gradients; and if those who are entrusted with the preparations of the time tables do not possess the technical information necessary to deal properly with the question of relative speeds, there is the strong probability that the programme prepared may be one both difficult and dangerous to fulfil. The spirit of rivalry is a strong incentive to fast running, but prudence and common sense should indicate that record speeds should only be attempted on the straight or favourable portions of the line. There is, unfortunately, the growing tendency to run faster and faster round the curved portion of our lines, heedless of the close approach to the limit of safety, and unless this excessive speed be controlled in time, the result must be disaster on a very large scale.

A sharp curve leading into or out of a terminal station or main-line stopping-station does not so much affect the train running as a sharp curve at an intermediate point between stations where the train may be expected to run at its maximum speed. Wherever it is possible it is very desirable to avoid sharp curves on inclines, because there are times when descending trains may acquire a considerable velocity, and wheels tightly gripped by the brakes have not the same facility for following the curves as when they are running free.

In rugged and mountainous districts sharp curves are almost unavoidable, except by introducing a series of tunnels; but in these districts both the gradients and curves are alike exceptional, the speed is necessarily slow, and special precautions are taken for the ascending and descending trains.

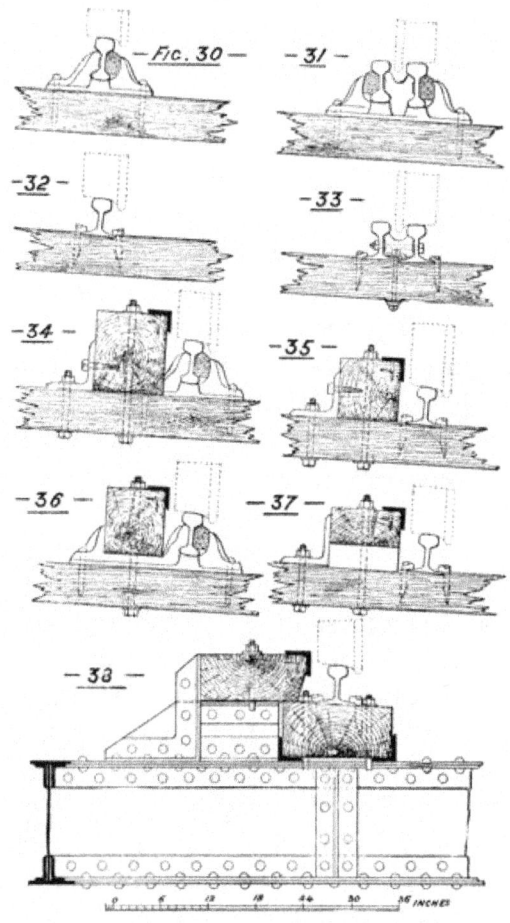

When setting out reverse curves on a main line a piece of straight line should always be laid in between the termination of the one curve and the beginning of the other, to allow of a proper adjustment of the rails to suit the super-elevation adopted on each of the adjoining curves. In station yards and sidings this is not so absolutely necessary, the sorting of the carriages and waggons and the marshalling of the trains being carried on at a low speed, which does not necessitate any super-elevation of the rails on the curves. The speed of the train regulates the amount of super-elevation to be given on any particular curve, and to ensure smooth and safe running this amount must be maintained uniform all round the curve. On curves of small radius, guard, or check, rails are frequently placed alongside the inner rail, as in Figs. 30 to 33, to check the tendency of the engine to leave the rails and run in a straight line. For the bull-head road a special chair is used, which holds both the running-rail and the check-rail, as shown on the sketch, the rails being kept the proper distance apart by the web portion in the centre, which forms part of the casting. For the flange railroad, check-rails are sometimes made of strong angle irons placed against the flange of the running-rail, and bolted to the transverse sleepers. This method is not nearly so strong or efficient as the arrangement shown on Fig. 33, with a cast-iron distance-block about six inches long, placed between the running-rail and check-rail, and all tied together with a strong through bolt. A bolt-hole is punched in the edge of the flange of check-rail, and a crab bolt and clip holds the two rails on the sleeper. The cast-iron distance-blocks are placed just outside the sleeper, so as not to interfere with the holding-down bolt. Doubtless these guard rails do good service, but if the leading wheels of the engine have sharp or worn flanges there is the possibility that the wheel, pressing against the high rail, may mount the rail, and throw the train off the line. A more secure method is to place the guard outside the high rail, as in Figs. 34 to 38. This can be done by securing a strong continuous longitudinal timber to the cross-sleepers —or to the cross-girders in the case of a girder bridge—with its outer or striking edge protected with a fairly heavy angle iron. The top of this outside guard above the rail level may be three inches or more, according to the height of any hanging spring, or portion of brake apparatus belonging to the rolling-stock. The distance between the striking-face of the guard and the inside of head of rail should be about 5 inches, or such width that before the flange of the wheel can mount on the top of the rail, the face of the wheel-tyre will be brought into contact with the striking-face of the outside guard, and thus effectually prevent the wheel leaving the rail. The sketches show some of the types applicable to the chair road, and to the flange railroad. In Figs. 34, 35, and 37, the outside brackets are of heavy angle iron cut off in lengths to correspond to the width of the sleeper. In Fig. 36 the cast-iron chair is lengthened, and has an end bracket to support the guard timber. In Fig. 37 a hard wood bolster is fastened on the top of each sleeper, and on this is placed the continuous guard timber. This method of increased security is frequently adopted on girder bridges and long iron viaducts which are on the straight, and in such cases it is usual to place the guards outside each of the rails forming the track.

The introduction of bogie engines and bogie carriages has conduced largely to the safe working of the train-service over the curved portions of many of our home railways, as well as to the economy in the wear and tear of permanent way and rolling-stock. The action of long rigid wheel-base vehicles passing round sharp curves is detrimental to all the parts brought into contact. Not only is there the constant tendency to mount the rails, and spread the gauge, but the tiny shreds of steel scattered all along close to the rail—particles ground off the rails, or off the wheel-tyres, or both—testify to useless wear, unnecessary friction, and great waste of motive-power.

The gradual increase of accommodation and conveniences in the carriage stock of European railways led to the gradual increase in the length of the vehicles. The six-wheeled carriage superseded the four-wheeled carriage, on account of its increased steadiness when running, but the introduction of long sleeping-cars, dining-cars, and corridor cars necessitated some better wheel arrangement than the ordinary six-wheel type could supply. The six wheels had been spread as far apart as was admissible for carrying weight and passing round curves, and something had to be done to meet the demand for still longer carriages. Many of the six-wheeled carriages at present running on our own home lines have a fixed wheel-base as long as 22 feet, and with this length the horn-plates must undergo a very considerable strain when adapting themselves for the passage round curves of small radius. On a curve of 15 chains radius (990 feet) a chord of 22 feet will have a versed sine or offset of $0 \cdot 73$ of an inch, and on a curve of 10 chains radius (660 feet) an offset of $1 \cdot 10$ of an inch. Fortunately, curves of the above small radius are not very numerous on our main lines; but wherever they do occur, the conflict between the long fixed wheel-base rolling-stock and the permanent way must be very severe to both. Several descriptions of eight-wheeled carriages have been tried on our home lines; but the system which is now most in favour is the ordinary bogie truck, which has been in use for so many years on all American railways. A bogie truck is really a short carriage frame complete in itself, with its wheels, springs, and brake appliances, and is attached to the under side of the carriage body by a central pivot, round which the truck can swivel or rotate sufficiently to adapt itself to the curved portions of the line. With a bogie truck at each end of a long carriage, the vehicle will pass as easily round curves as on the straight line, side pressure, or grinding against the rails, is obviated, and friction is reduced to a minimum. The bogie truck may consist of four wheels or six wheels, according to the length and weight of the carriage to be supported.

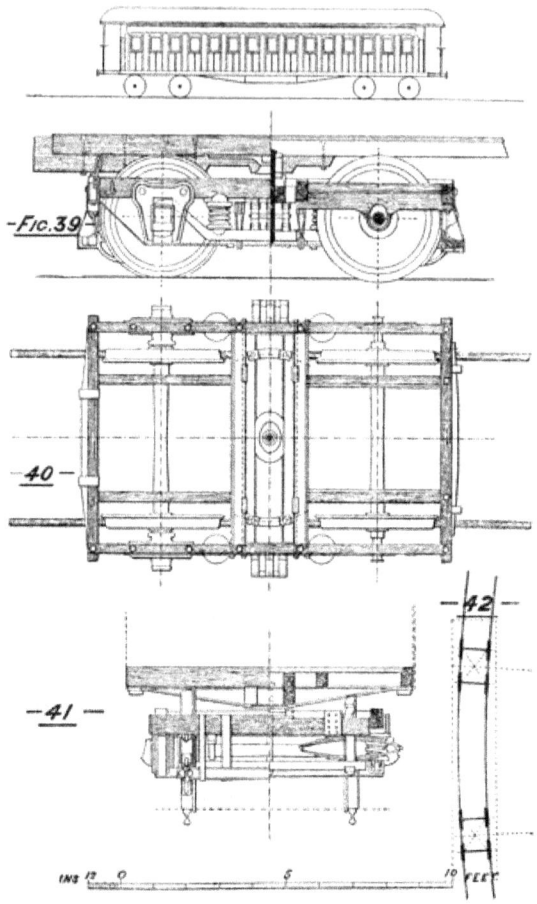

Figs. 39, 40, and 41 show sketch elevation, plan, and transverse section of one pattern of four-wheel bogie truck largely adopted in American carriage stock, and although there are other types varying in detail, the general principle remains the same in all. The diagram sketch (Fig. 42) represents the two bogie trucks slightly swivelled to adapt themselves to the curve round which the carriage is supposed to be passing.

For carriage or waggon stock with an independent bogie truck at each end, the central pivot and swivelling motion supply all the freedom that is requisite; but for locomotives it is necessary to provide for lateral as well as for swivelling movement. The driving and trailing wheels—and sometimes one or two other pairs of wheels—work rigidly in the frames, and as the normal position of the centre of the bogie truck must be in the centre line of the engine for the straight line, it is evident that some appliance must be introduced to allow the truck to move laterally when the engine has to traverse the curves.

Figs. 43, 44, and 45 give sketch elevation, plan, and transverse section of a swing-link bogie truck as applied to an ordinary American locomotive. Its recommendations are its simplicity, its efficiency, and its accessibility for inspection and lubrication. The swing-links, which provide for the lateral movement, are direct acting, and do not require any side springs of steel or indiarubber. All the principal parts of the bogie are visible and not mysteriously cased in with plate-iron boxwork.

In the sketches several minor details are purposely omitted and only sufficient particulars shown to explain the method of working. The under side of the upper centre plate which carries the cylinder castings and smoke-box end of boiler is cup-shaped, and fits into an annular groove or channel in the lower centre plate, which is suspended from the framework of the truck by the four swinging links. Practically the entire carrying and swivelling work of the bogie truck is effected by the annular-groove casting moving round the cup-shaped casting, and the centre pin is merely passed down through each to guard against the risk of the one lifting out of the other from sudden shock or derailment.

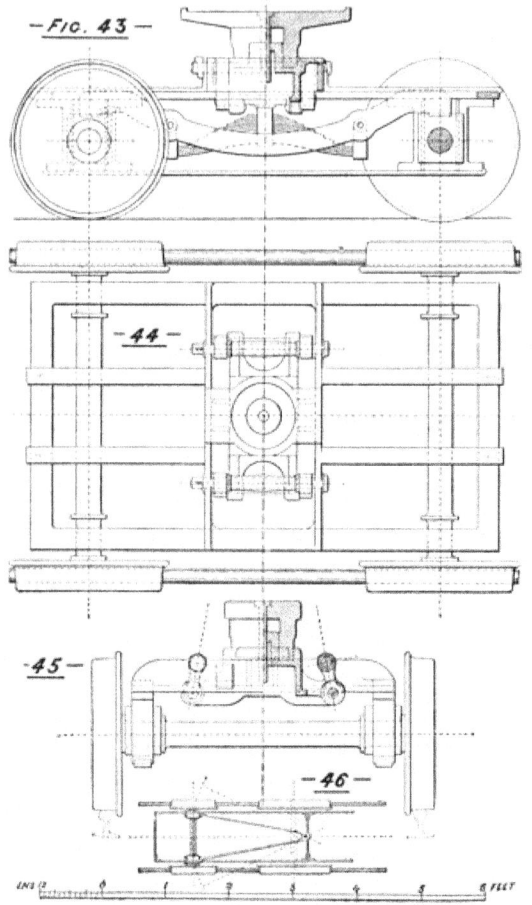

The lateral movement of the truck is obtained by means of the four swing-links. When the engine is on the straight road the centre line of the bogie is on the centre line of the engine, and the links hang in the positions shown on the sketch, inclined towards the centre; but upon entering a curve they come into play, and allow the truck to move out sideways to the right or left, according to the direction of the curve, the one pair of links assuming a flatter angle, while the other pair approach nearer to the vertical, the extent of side movement depending on the amount of the curvature. When the engine enters the straight line again, the bogie truck resumes its central position.

The Bissell truck consists of one pair of wheels connected to a triangular framework, as shown in Fig. 46. The axle-boxes are attached to the side of the triangle which lies parallel to the axle, the other two sides terminate in a circular ring which works round a centre pin fixed to the engine. These two sides are practically the radii of a given circle, and permit a large amount of lateral movement, which can be controlled by placing suitable stop-pieces to limit the side play to the extent desired.

Radial axle-boxes have been tried on the engines of some railways. In the best types the opposite boxes are braced together by a diaphragm, or plate-iron framework, to ensure that both boxes work together. The curved faces of the horn-blocks, in which the radial axle-boxes slide, are struck from a centre taken at some point to the rear of the normal centre line of the axle, and stops are placed at proper distances to control the extent of lateral movement. Although the advocates of radial axle-boxes may urge some points in their favour, there are few engineers, if any, amongst those who have had practical experience of both systems, who would for a moment claim for the radial axle-box anything but a modicum of the many advantages obtained by the four-wheeled bogie truck.

As one of the principal functions of a four-wheeled bogie truck for an engine is to act as a path-finder, or guide, to the other wheels which constitute the fixed or rigid wheel-base portion of the machine, it follows, therefore, that the full benefit of the bogie truck can only be obtained when it is placed at the leading, or front, end of the engine. In this position the bogie, with its swivelling arrangement and smaller weights, is the first to pass over the rails, and in doing so shapes the course and prepares the way for the easy running of the heavier wheel weights which have to follow. When the bogie truck is placed at the rear end of the engine, its action is restricted to affording lateral movement only, and the driving and coupled wheels have to force or pound themselves round the curves in a jerky, irregular manner, as compared to their smooth running when following the leading or guiding influence of a bogie truck in front.

The wheel-base of a four-wheeled bogie truck for an engine should always be greater than the gauge of the line over which the bogie has to travel On the 4 feet 8½ inch gauge some of the best results have been obtained with bogies having wheel-bases varying from 6 feet to 7 feet. Where the wheel-centres have been less than 6 feet, the running has been found to be much less steady than with the wider spacing; and where the wheel-base is not more than the gauge, there is a tendency for the bogie to catch, or lock, when passing round sharp curves.

back to form [Transcriber's Note: Searchable text of form]

ESTIMATE OF THE PROPOSED (RAILWAY).

Line No.
Length of Line:
Miles. F. Chs.
Whether single or double.
Cubic yards. Price per yard.
Earthworks:
Cuttings—Rock Soft soil Roads Total
Embankments, including roads, __ cubic yards
Bridges, public roads—number
Accommodation bridges and works
Viaducts
Culverts and drains
Metallings of roads and level crossings
Gatekeepers' houses at level crossings
Permanent way, including fencing:
Miles. F. Chs. at
Cost per mile.
Permanent way for sidings, and cost of junctions
Stations
Contingencies __ per cent.
Land and buildings
Total £
Dated this   day of   18__
Witness:
Engineer.

CHAPTER II.

Works of construction: Earthworks, Culverts, Bridges, Foundations, Screw piles, Cylinders, Caissons, Retaining walls, and Tunnels.

Earthworks.—Under this heading may be classified cuttings and embankments of earth, clay, gravel, and rock.

When setting out a line and adjusting the gradients, an endeavour is usually made to so balance the earthworks that the amount obtained from the cuttings may be sufficient to form the embankments. With care, this may be effected to a considerable extent; but there will be places where the material from cutting is unavoidably in excess, and others where the cuttings are too small, or contain good rock, or gravel, which can be more advantageously used for building and ballasting purposes than for ordinary embankment filling. Or there may be a large cutting which will provide enough material to form three or four of the adjoining embankments; but the distance, or *lead*, as it is termed, to the far embankment may be so long, and, perhaps, on a rising gradient, that it would be cheaper to run the surplus cutting to *spoil*, and *borrow* other material for the far embankment from side cutting or elsewhere. A long lead adds materially to the cost and time of forming an embankment, as it not only necessitates a considerable length of *service*, or temporary permanent way, but also occupies much time in the haulage of the earth waggons. For distances of half a mile and upwards, a small locomotive is more suitable than horses for conveying the waggons.

To run to *spoil* is the term applied to such of the material from a cutting which, not being required or utilized in the formation of the line embankments, is removed and tipped into mounds, or *spoil-banks*, in some one or more convenient sites near the mouth of the cutting. Sometimes the surplus material is disposed of by increasing the width of the embankments. Material excavated in a tunnel, and hoisted through the shafts to the upper surface, has to be deposited in spoil-banks along the centre line of the tunnel.

To *borrow* material to form an embankment is the term used when the earthwork filling is not obtained from the cuttings on the line. This borrowing is generally done by excavating a trench on each side of the line, of such width and depth as will supply sufficient material to form the embankment. Fig. 47 gives an example of an embankment thus made from side cutting. In some cases a piece of high ground adjacent to the embankment can be utilized for obtaining a portion, or even the whole of the filling.

Increased material is sometimes obtained by widening the cutting, or flattening the slopes, or both.

The degree of slope of a railway cutting must be regulated by the nature of the material excavated. A slope of 1½ to 1, which gives for every foot of vertical height a width of one foot 6 inches of horizontal base, as in Fig. 48, is usually adopted for cuttings in ordinary earth, good clay, sand, or gravel. There are some descriptions of strong clay and marl which will stand at a steeper slope, even at 1 to 1; but, on the other hand, there are some kinds of clay which must ultimately be taken out to 2 to 1, and even 3 to 1.

It frequently occurs that the slopes of a clay cutting, taken out to 1½ to 1, appear to stand well for a time, but after exposure to the frost and rain of one or two seasons, the material becomes loosened, and forms into slipping masses, which slide down on to the line, stopping all traffic, and have to be cleared away before train operations can be resumed.

Cuttings through solid rock may be taken out to a slope of ¼ to 1, as shown in Fig. 49, provided the material is compact, and there is not too great a dip in the strata or rock-beds. Where the rock-beds lie at a considerable angle, the slope on the high side will have to be made flatter than the slope required on the low side, as shown in Fig. 50, and great care must be taken to remove from the high side all loose or disconnected pieces of rock which might come away and slide down on to the line.

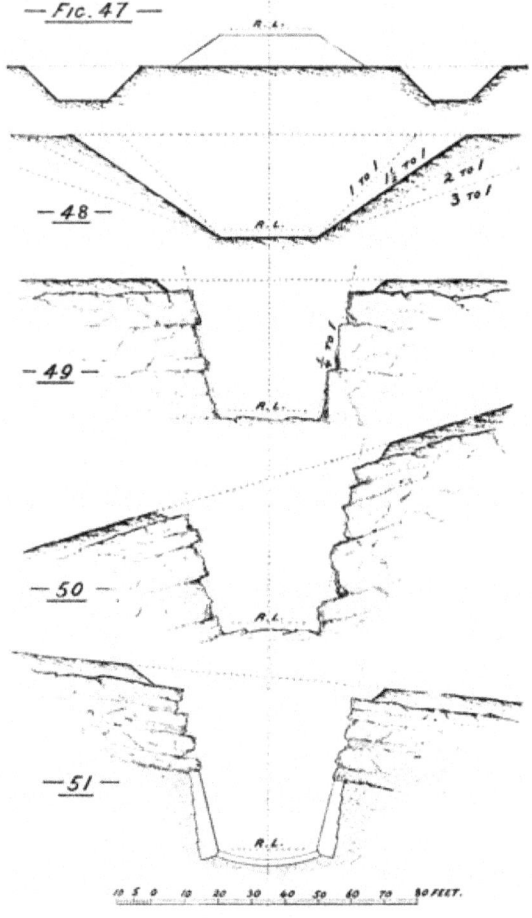

Strong dry chalk will generally stand at a slope of ⅓, or ½ to 1, but when wet and mixed with flints it will be necessary to increase the slope to not less than ¾ to 1. Where the rock is loose and disintegrated, a slope of not less than ½ or ¾ to 1 will be required, and at many points there will be detached threatening masses of rotten rock which must be cleared away to a much flatter slope for safety. In cuttings of this description it is frequently found necessary to clear out a portion of the loose pieces of the lower cavities and build in their place a facework of masonry to support the superincumbent rock. Springs of water rising in the rock, or running over any part of the rock slopes, must be properly provided for, and conducted to the nearest channel. They should be carefully watched during the winter season, when the frost, acting on the water penetrating the crevices, splits and separates large pieces which were previously firm and secure.

Instances will occur where a cutting has to be made through a thick bed of rock and several feet of soft loose strata underneath. The effect of forming a cutting through the soft strata is to induce the heavy bed of rock above to squeeze or force out the softer material below, and unless proper means were taken to avert such a disturbance, the entire cutting would have to be excavated to a very flat slope. The method adopted in such a case is to build strong face-walls of masonry, brickwork, or concrete, underneath the rock, as shown in Fig. 51, with strong inverts placed at short distances. Suitable arrangements must be made to take away the drainage water which will collect at the back of the walls, and weeping-holes or outlets must be left in the lower part of the walls to convey the water into the water-tables on the line.

Where there is a depth of earth cutting on the top of the rock, the earth should be cut away so as to leave a bench or space of 3 or 4 feet between the edge of the rock cutting and the foot of the earth slopes, as shown on Fig. 52.

In cases of shelving rock, with earth or clay on the top, as shown in Fig. 53, it is frequently found necessary to remove the whole of the clay on the high side to prevent the possibility of its sliding off the rock on to the line below.

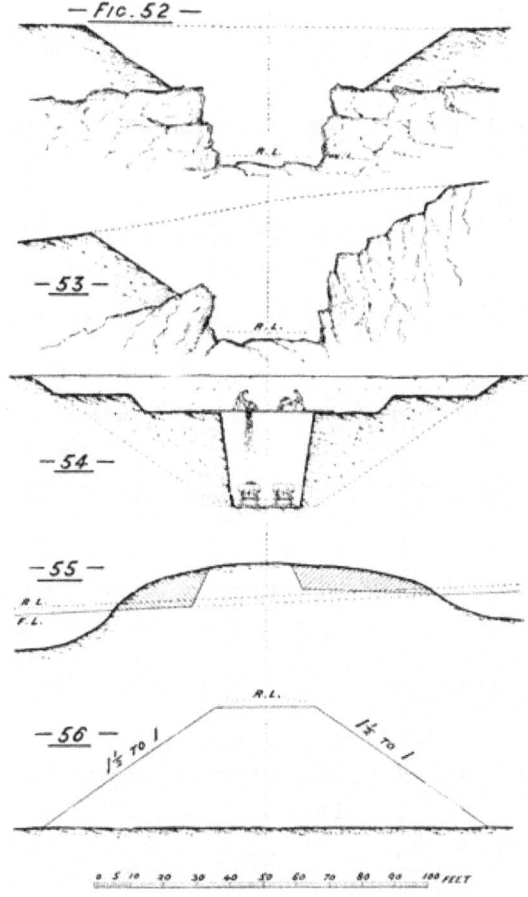

In large cuttings it is usual to push forward a gullet of sufficient width for one or two lines of waggons, as shown in Fig. 54. When this has advanced some distance, strong planks or half balks of timber are placed across the gullet, and the sides or wings of the cutting can be excavated, the material wheeled to the gullet, and tipped from the barrows into the waggons beneath. By this arrangement the work can be carried on very expeditiously, as one set of men can be engaged advancing the gullet and laying the track, while others are following up and taking down the sides. A large number of waggons can thus be filled in a day, and a small locomotive kept fully employed.

Occasions will arise where the material from a large cutting, situate on a continuous gradient, as in Fig. 55, has to be carried in both directions to embankment.

In wet weather, or if the cutting is at all wet, it would be almost, if not quite, impossible to carry on the excavation at the upper end to the proper formation level. The water would collect at the lower level, and not having any means of escape, except by pumping, would stop the work. In such a case the best way is to take out the cutting at the upper end to a slight rising gradient, as shown in the sketch, sufficient to carry away all water, and afterwards take out the lower portion in the working from the other end of the cutting.

Cases will arise where it will be necessary to make a shallow cutting through boggy peaty ground. If the boggy material be very soft, and its thickness from the formation level to the solid ground below be not great, it may be advisable to remove this extra thickness down to the hard lower bed, and fill in up to formation level with strong material. If, however, the bog or peat be too thick to justify its entire removal, it should be excavated say down to two feet below formation level, and a thick layer of branches of trees and strong brushwood closely laid and packed the full width of the road-bed. On this preparatory foundation must be placed good clean ballast to carry the permanent way. Two or three extra sleepers should be allowed to the rail length, and in some instances it will be necessary to introduce two, or even four, rows of strong longitudinal timbers—half balks—under the transverse sleepers. The object of all this extra timber is to obtain a large increase of bearing area on the soft yielding surface of the boggy material. Notwithstanding these special precautions, the trackway will sink down a little during the passage of an engine or train, but will generally return to its former level. Good side drains or water-tables should be formed at each side of the cutting to take away all rain and surface water.

In all cuttings it is desirable to have the line of formation on a slight gradient, sufficient to carry away all rain water or spring water which may be collected in the water-tables; but more particularly so is this necessary in a rock cutting, where the material, being non-absorbent as compared with earth or gravel, requires that all drainage must be carried away to the mouth of the cutting.

In carrying out railway embankments and road approaches, it is usual to form the sides to a slope of 1½ to 1, as shown on Fig. 56.

Occasionally the cuttings produce material which might stand at a rather steeper slope, but considering the effects which might afterwards be produced by heavy rains falling on the sides, it is more prudent to adopt the flatter slope of 1½ to 1. Some descriptions of clay will not stand at the above slope, but require a slope of 2 to 1, or even 3 to 1.

When proceeding with the earthworks, it is customary to first remove and lay aside a layer, say 9 inches in depth, of soil and earth from the seat of the embankments and top widths of the cuttings, to be used afterwards in soiling the trimmed and finished slopes of the cuttings and embankments. This soil being removed, the actual work of the excavation can be commenced. The working longitudinal section will give all the necessary particulars as to position of the mouths of the cuttings and the depths at the various chain-pegs, and the top widths of the cuttings can be ascertained by calculation, if on even ground, or from the cross-sections if on side-lying ground, according as the material may be earth, clay, or rock.

For facility of carrying on the works, reliable bench marks, or reduced level stations, must be established at convenient distances along the route of the line, and from these and the fixed chain-pegs the correct line of formation level can be checked from time to time as the work proceeds.

For ordinary earth or clay cuttings, the usual tools are picks and iron crow-bars for loosening, or *getting* the material, and shovels for filling into barrows, carts, or waggons. For heavy earthworks, steam excavators are now largely employed. Great improvements have been made in this class of machinery, in the way of perfecting the method of excavating lifting, and filling the material into the earth-waggons.

In nearly all rock cuttings the greater portion of the material has to be taken out, or loosened, by blasting with gunpowder, dynamite, or other explosive. The number and extent of the charges will depend upon the nature of the rock and its stratification, and also on its position as regards proximity to buildings or residential property.

Where the rock is loose, or disintegrated, the pieces can generally be readily separated by picks and bars without having to resort to any great extent of blasting.

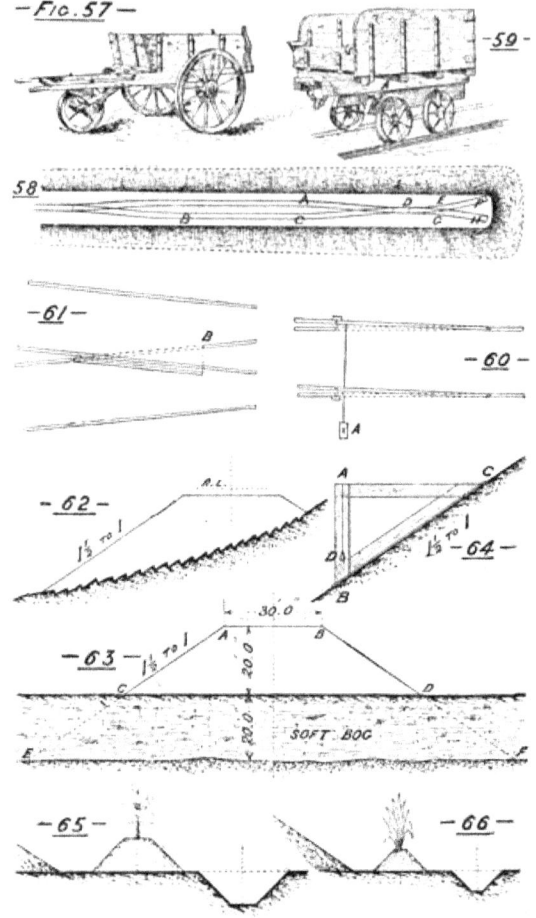

The first of the material excavated in the cuttings is generally conveyed in wheelbarrows to form the commencement of the adjoining embankments. When the wheeling distance becomes too far for economical barrow work, ordinary carts or three-wheeled carts, sometimes termed *dobbin carts*, are brought into operation where the cuttings and embankments are light; but where the earthwork is heavy, both in excavation and filling, a service or temporary road of light rails and sleepers is usually laid down to carry strong *tip* earth-waggons. For moderate distances these waggons are hauled by horses, but for distances over three-eighths of a mile a small locomotive is more speedy and economical. Fig. 57 shows one form of dobbin cart; the wheels are made with good broad tyres, so as not to sink too deep into the soft ground, and the body being attached to the framework by a pivot or trunnion on each side, can be readily tilted over, and the earth tipped out, by releasing the holding-down catch. Where the ground is soft and wet, or of a very loose sandy nature, the work of hauling these dobbin carts is very heavy on the horses, and in such cases it soon becomes an advantage to lay down a service road of rails and sleepers. This service road is formed of light rails manufactured for the purpose, or old, worn rails no longer fit for main-line work, spiked down on to rough transverse wooden sleepers. The end of the embankment in course of formation, and where the earth is being tipped, is termed the *tip head*. Two or more roads are required at the tip head to form the embankment to its full width. Fig. 58 gives a sketch plan of a service road near the tip head. The width is shown as for a double line. The earth-waggons are hauled along the line from the excavation, and brought to a stand at the point **A**. If a locomotive has drawn the waggons, it is then detached, moved forward, and shunted back into the siding **BC**. A horse accustomed to tipping then takes one full waggon at a time over one or other of the two turn-outs, **DEF** or **DGH**, to the tip head, sufficient impetus being given to the waggon to run the front wheels off the ends of rails on to cross-sleepers laid close, with a steep rise, and backed up with earth. This suddenly checks the frame of the waggon, and the body containing the excavated material revolves on its trunnion, tilts up, and shoots out the material well forward, so that the man in charge of the tip head, who also knocks up the "tail-board catch," is able to level off the filling without assistance. The empty waggon is then hauled back, and turned into the siding **BC**, and another full waggon taken forward and tipped, until all the waggons of the rake are emptied. Ten waggons generally form a rake when the work is pushed forward vigorously, each waggon holding about three tons. The tip head horse pulls the waggon by a trace-chain having a spring catch at the end, by which the driver releases the horse at the right moment. It is very important that this spring catch should be kept in good order, because occasionally too much impetus is given to a waggon, which, running over the tip head down the slope, would drag the horse with it if the spring catch did not act properly. Good firm foothold must be provided for the tipping horse.

The tip head should never be carried across culverts or bridges until they have been well backed up, and protected by a thick covering of earth or clay, wheeled in with barrows to an equal height on each side of the masonry, so as to prevent undue side pressure.

Fig. 59 gives a sketch of one form of end-tipping waggon. In some cases the wheels are made of cast-iron, but as these are readily broken during the rough handling to which earth waggons are exposed, it is questionable whether the light wrought-iron wheels, with light steel tyres, used on some works, are not more economical in the long run. The framework and body are made of strong undressed timber, well bound and bolted together. The tail-board catch keeps the body of the waggon in its proper horizontal position while loading or running, but when released leaves the body free to tilt up, and to revolve on the front trunnion by means of the circular clip **A**. The same principle is also applied to side-tipping waggons which are used for the widening of embankments, or formation of platforms and loading-banks.

The permanent way of these service roads is generally made as simple as possible. A pair of movable rails are used instead of switches, as shown in Fig. 60. These rails are linked together by iron tie-rods, and pulled or pushed over into position for one or other of the roads by means of the handle at **A**. A stout iron pin, or iron clamping-plate, serves to retain the rails in position during the passing of the waggons. In a similar manner, a short rail working on a pin, or pivot, is made to answer the purpose of an ordinary crossing. The rails are laid complete and continuous for the one road, and for the second road the outer rail is laid sufficiently high to cross over the rail of the first road. A piece of rail is then secured by a centre pin, or pivot, to the cross-sleeper, as shown on Fig. 61. This pivoted rail is pulled over into the position shown by the dotted lines, to allow the passage of waggons on the one road, or pulled across to the end of rail at **B**, for waggons to pass on or off the other road. In the latter case an iron pin or clamp serves to keep the pivoted rail in position. As these service roads are merely laid down on the soft loose material brought forward for filling, they require constant packing and lifting to prevent them working into depressions, which might cause the waggons to leave the rails.

To indicate the height of the embankment filling, strong stakes or poles must be firmly set in the ground at each chain-peg. On each of these poles two cross-bars must be fixed, the lower one placed to the correct height of the embankment, and the upper one to show the amount allowed for subsidence. The excavated material, as brought from the cuttings, is in a soft, loose condition, and an allowance must be made for its settlement, or subsidence, as the embankment becomes consolidated. This allowance will, of course, depend on the height of the embankment and the quality of the material, but for ordinary earth and clay it is customary to allow about one inch to the foot of height, which is equal to about 8 per cent.

When forming embankments over very side-lying ground, it is necessary to cut steps in the sloping surface on which the filling material has to be placed, as shown in Fig. 62. These steps give a hold to the new earthwork, and check the tendency to slide down the hillside.

Embankments have frequently to be carried over ground which is low, soft, and wet, but not boggy. If the culverts and drains are sufficiently large, and properly arranged, these places are not likely to cause much future trouble.

For a thoroughly soft deep bog, however, it is most difficult to make any accurate calculation as to the amount of embankment filling which will be necessary to form a permanent foundation for the line; and the construction of a high heavy embankment across such a place is one of those undertakings which every engineer is most anxious to avoid. A large quantity of material may be tipped into the bog, and seem to stand fairly well for a time, and then suddenly disappear altogether. More material has to be brought forward, and will most likely disappear in a similar manner. The filling material being heavier than the bog on to which it is thrown, falls through, and displacing the soft semi-liquid matter, continues to sink down lower and lower until it is stopped by a harder stratum underneath. In a measure the operation somewhat resembles the tipping of earth into a lake; the material will go down until it meets with a solid bottom, and in going down it assumes its own natural slope, and forms for itself a width of base corresponding to its height. It will be readily understood what an enormous amount of filling material will be swallowed up in following out such a process. On a very soft bog, say 20 feet in depth, over which an embankment 20 feet high has to be formed, the extent of the actual earthwork filling will very probably closely approach the outline shown in Fig. 63. The upper portion, **ABCD**, representing the embankment proper, will contain about 133 cube yards to the yard forward, whereas the lower portion, **CDEF**, which has displaced the soft boggy matter, will contain about 266 cube yards to the yard forward, or, in other words, the filling which is out of sight will be double the filling which is in view above the section ground line.

Apart from the large amount of filling consumed in forming this semi-artificial island, the progress of the work itself is very perplexing. A long length of the bank may have been raised again, once or twice, to the proper height, and may have carried rails and earth-waggons for some weeks, and then sink all at once several feet. The sinking, too, may not be uniform, but may produce fissures, depressions, and separation of the earthwork which will necessitate much care when bringing forward fresh filling material. The bog may not be of the same consistency throughout, there may be some layers of harder material, such as imbedded trunks of trees, and these may sustain the filling for a time, and then yield under the increasing weight of the superincumbent mass. Even when the embankment is finished throughout, and shows no sign of sinking, it should be very carefully watched for a long time for any indication of further movement.

When the bulk of the material has been taken out of an earth or clay cutting, the work of trimming the slopes should be put in hand, so that any surplus left on the wings, or sides, may be removed, and carried away before stopping the earth-waggons. The angle of slope having been decided, a battering rule of light wooden boards is made to correspond to the slope, and in form similar to that shown in Fig. 64. A plumb-bob is suspended from a fixed point, **A**; the lower end, **B**, is then held against a peg or mark which indicates the correct level and width of the cutting at the place, and the upper end, **C**, is raised or lowered until the plumb-bob string coincides with the vertical line marked on the rule from **A** to **D**, and the plumb-bob rests steadily in the space cut for it at **D**. With this battering rule a length of seven or eight feet, according to the size of the rule, is first trimmed to the correct slope, and by continuing the application of the rule up the side, a correct slope line is obtained from bottom to top of slope at that place. By repeating the process at convenient distances along the cutting, a series of correct slope lines are obtained, and the intermediate space can readily be trimmed to correspond.

The same form of battering rule and method of working is applicable for trimming the slopes of the embankments.

When the slopes of the cuttings and embankments have been trimmed, vegetable soil, which has been laid aside, or reserved as previously described, should then be spread evenly over the slopes to the uniform thickness of not less than four inches, and the whole sown with good grass seeds to form a strong sward.

The trimming, soiling, and sowing of the slopes not only gives a more finished appearance to the earthworks, but the strong grass, when once well grown, binds the surface together, and helps to resist the injurious effects of heavy rains and melting snow.

There are many places abroad where a neat finish to the earthworks is considered quite a secondary matter, or where it would be difficult to obtain suitable soil to spread on the slopes. The earthworks are hurried forward to allow the iron highway to be laid down as quickly as possible, the slopes of the cuttings and embankments are only roughly trimmed, and nature is left to supply such grass or vegetation as may spring up, or be self-sown.

The fencing in of a line of railway serves the double purpose of defining the boundary of the company's property, and of forming a barrier for the prevention of trespass of persons and animals on to the line. For our home lines, fencing is compulsory, and the same obligation exists on many foreign railways. In our colonies, and out in the far West of the United States, and in newly opened out countries, fencing, except near towns and villages, is rather the exception than the rule; people and animals roam at will from one side of the railway to the other wherever they find a convenient crossing place, and the cowcatcher of the engine has to be depended upon for throwing aside any animal which may be standing, or resting, on the line of rails at the passing of a train.

The description of fence will be influenced by the locality, and the materials conveniently obtainable. Where stone is plentiful, perhaps brought forward out of the cuttings, and labour cheap, a masonry wall will be found a most suitable permanent fence. Any fence to be of service should not be less than four feet high. A wooden post and rail fence is much in favour in some districts, the posts being firmly set or driven into the ground, and four or five stout bars nailed on to, or set into, the upright posts. This fencing does not last very long, the pieces are small in size, and soon fail from decay. Quick or hawthorn hedges, when fully grown, make a good fence, but require careful attention to prevent gaps being made by roving cattle. They also require constant trimming and cutting. The quicks are generally planted in a mound formed by cutting a continuous ditch, or gripe, as shown in Fig. 65. The ditch serves as a drain to take away water running down the slopes of the embankments, small openings in the mounds, or drain pipes through them, forming leaders to conduct the water to the ditch or gripe. The outer edge of the ditch represents the boundary of the railway property, unless specially arranged otherwise.

Galvanized iron-steel wire fencing, if not made too light, is strong and durable, and very easily kept in order.

The wires may be secured to strong wooden posts, which should be creosoted, and not placed too far apart, or to iron posts or standards of angle iron or tee-iron section. The straining-posts, whether of iron or timber, must be stronger than the intermediate posts, firmly fixed into the ground, and well stayed, to withstand the pulling and tightening of the wires. There are many places where a quick fence would not grow, and where the ground is too soft to carry a wall. In such cases a good galvanized-wire fencing will fulfil all requirements. The strand wire is better than the plain wire, as its method of manufacture necessitates the use of a superior material, and it is easier to straighten and keep in good order. An extra strong fence is often made of six, eight, or more rows of round rod-iron secured to wrought-iron uprights of bar-iron or tee-iron.

In hot countries abroad an excellent fence is obtained by planting a species of cactus or aloe in a similar manner to the quick fences at home, and as shown in Fig. 66. These cactus plants are readily obtained, are very hardy and quick in growth, and with their large spike-shaped leaves form such an almost impenetrable barrier that few animals will attempt to pass.

Road approaches to bridges over or under the line, or to public road level crossings, may be fenced in the same manner as the line proper. If quicks are adopted, it will be necessary to put up a light wooden fence also to protect the young plants until they are well grown. Near towns and villages it is frequently found advisable to adopt a specially strong wooden fence, or close-boarded fence, where the approach is an embankment, and too newly made to carry a wall.

Gates for farm or occupation level crossings may be made of wood or iron. As a rule, iron gates are preferred, as they can be supplied at the same cost as wood, and are very much more durable. Gates for public road level crossings have to be so placed that they will either close across the railway or across the road; their length will therefore depend upon the width and angle of the road crossing. It is better to make these gates of wood, so that, in the event of a train running through them, there may be less risk of injury to life and rolling-stock than if they were made of iron. For footpath crossings, small gates, wickets, or stiles may be adopted of such form as may be found most suitable for the requirements.

Culverts and Drains.—Before proceeding with the formation of the embankments, it is necessary to construct the culverts and drains which will be covered over by the earthworks. Any existing drains which may be of too light a description must be reconstructed in a more substantial manner. It is a simple and comparatively inexpensive matter to rebuild a drain before the earth filling is brought forward, but it is a costly work to open out an embankment, and rebuild a culvert afterwards. Unless the seat of an embankment is well drained and kept free from the accumulation of running water, the earthwork will be exposed to washing away of the lower layers, and consequent subsidence. Each watercourse or open drain must be provided for either by a separate culvert of suitable size or, as may be done in some cases, by leading two or more watercourses into one, and thus passing all through one culvert of ample capacity. When fixing the sizes of the culverts they must not be limited to the normal flow of water, but a large margin must be allowed sufficient to meet extraordinary floods. The depth of the bed or invert of a culvert is a very important point. If laid too high, and the stream above should at any time deepen, the high invert would check the flow of the water, and would also incur the risk of being undermined and gradually carried away. If, on the other hand, the invert be laid too low, it will gradually silt up to the level of the stream-bed alongside, and there will be so much of the culvert space lost for all practical purposes. In cases when the invert of a culvert has to be laid at a special low depth to allow for future improvements in drainage, it is advisable to give extra height from the invert to the crown, or top, so as to provide ample waterway in the event of any silting up in the mean time. Particular care should be taken when building the foundation of a culvert. It has to be laid on the site of the watercourse, or on a new channel which will ultimately form the watercourse, and it should be built sufficiently deep into the ground to avert as far as possible the chance of water finding a course through below the foundation.

The invert may be of stone pitching or brick if the current is not rapid, or liable to bring down stone boulders from its gravelly bed.

With a stream-course having considerable fall, and which carries with it large stones, roots of trees, and other *débris*, the invert should consist of strong pitching, composed of large-sized, rough-dressed stones of hard, durable quality, capable of withstanding the pounding of the boulders brought down during floods. A soft description of stone would be quite unsuitable for the invert of such a stream; the pitching would wear away quickly, break, and become detached, leaving the foundation and side walls exposed to the cutting inroads of the water.

Where large flat bedded stones or flags of tough quality can be obtained, they form good covers, or tops, for culverts up to two feet in width. They should have not less than nine inches bearing on the side walls, and their contact edges should be fairly dressed, so as to fit sufficiently close to prevent the embankment filling from falling through.

Where the stream, or run of water, is very small, strong earthenware pipes, 9 inches or 12 inches in diameter, well bedded, may be sufficient to carry away all the water likely to arise. For small springs in low swampy ground, dry stone drains may in many cases be used with advantage. These are made by cutting a trench, say two feet deep by twelve or eighteen inches wide, in the seat of the embankment from side to side, and filling it up with dry rubble stones, not boulders, hand-laid, the upper layer placed on the flat to keep the earthwork as much as possible from filling in between the stones.

In soft boggy ground, where the depth to a hard bottom is very considerable, wooden culverts are frequently adopted. Although these cannot be classed as permanent structures, still, when they are made of sound well-creosoted timber, and substantially put together, they last for a number of years. Sometimes they are made cylindrical in section—a species of elongated cask with strong iron hoops every few feet. Others are rectangular in section, made with two strongly trussed side frames connected and covered with cross-planking and longitudinal tie-planking on the top and bottom.

Wooden culverts are seldom made of very large size, rarely exceeding an opening of 3 feet, and it is considered preferable to use two of these culverts of moderate dimensions than one of large size. Figs. 67 and 68 give sketches of wooden culverts of cylindrical and rectangular section, and Fig. 69 of flag top culverts of 12-inch, 18-inch, and 2-foot openings. In masonry culverts the side walls are shown to be of rubble stonework, but brickwork can be used instead, provided the bricks are well burnt, hard, and capable of withstanding the action of the water.

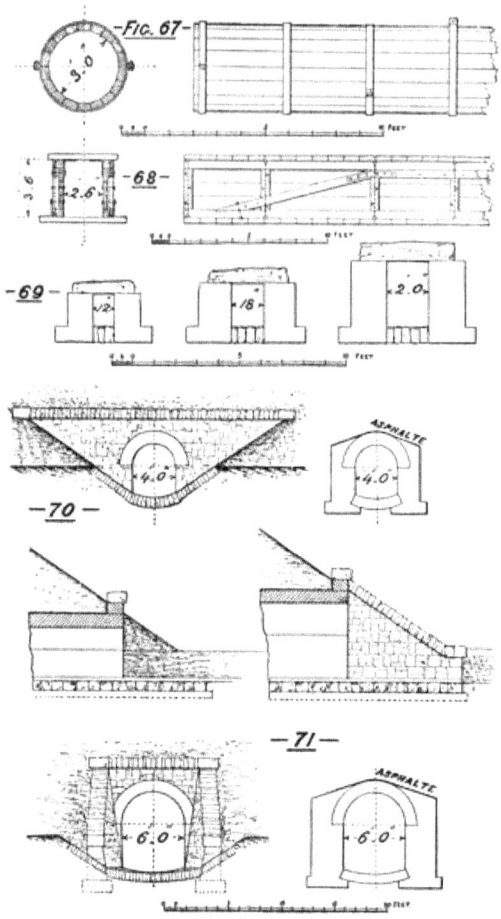

In Figs. 70 and 71 are shown types of arch-top culverts of 4 feet and 6 feet span respectively. The arch portion is shown to be of brick, which, as a rule, is cheaper than stone rings, which must be cut and dressed to suit the small radius of the arch. The side walls may be of brick of good quality. Occasionally they are built of concrete. The wing walls may either be carried out in the direction of the stream, as in the sketch of the 6-foot culvert, or they may be built transverse, as shown on the 4-foot culvert, whichever arrangement is found to work in the best for the case in question.

For arch culverts on very steep side-lying ground it is better to build the arch-top in steps, as shown in Fig. 72, instead of forming it parallel to the invert, or slope, of the stream-course. The level portions of the arching give a better hold for the embankment than could be obtained on a long inclined surface of brickwork or masonry.

The writer has built a large number of culverts of this type for mountain streams on steep hillsides, and has found them to prove satisfactory in every way.

In embankments alongside tidal rivers, or across the corners of estuaries of the sea, culverts have frequently to be so constructed that they will permit the passage of the drainage water from the land, or high side, without admitting the tidal water. This can be arranged by placing at the lower end of the culvert close-fitting hinged-flap valves opening outwards. When the tide has gone down the weight of the fresh, or land, water swings the flap-valve sufficiently open to allow of a free passage; and, on the other hand, when the tide rises, the pressure of the water against the face of the flap-valve keeps it tightly closed, and prevents ingress of the salt water.

Culverts are sometimes fitted with lifting-valves or doors, which can be raised or lowered to serve irrigation purposes. The door, which works in guides, is made sufficiently heavy to fall with its own weight, and the raising is effected by means of a screwed suspension-rod working in a well-secured fixed nut.

In cases of soft or treacherous ground, timber-piling or wide bed-courses of cement concrete are necessary to form firm foundations for culverts. Drains and streams which are intersected by a railway cutting have to be dealt with according to their size and their height above the finished rail level. The water from a small drain or field spring may be conducted in pipes down the slope of the cutting into the water-table, or side drain, at formation level, and will be thus carried away to the lower level at the entrance of the cutting. In many cases streams can be diverted, and the water led away to some lower point without the necessity of actually crossing the railway. With a large stream, where it is essential that the water should be conveyed across the line and continue on its ordinary course, it may be carried over in iron pipes or iron trough if there is ample headway, or in iron syphon pipes where the height is not sufficient. The iron pipes or trough can be supported on masonry or brick piers, or cast-iron columns, the height from the rails to the underside of the conduit being not less than that adopted for the over-line bridges.

Occasionally the pipes can be carried across on an over-line bridge, either by placing them under the roadway or on small brackets outside the parapet.

With the syphon arrangement the iron pipes must be laid down the slopes of the cutting and under the road-bed of the permanent way. The pipes must be continuous, strong, and firmly connected at the joints to prevent leakage. The inlet and outlet ends of the pipes should be securely built into receiving-tanks of masonry, brickwork, or concrete, to ensure an uninterrupted flow of the stream, and also to prevent any of the water from percolating through under the pipes and on to the railway. As a precautionary measure, it is well to place iron gratings some little distance in advance of the syphon pipes to intercept and collect any brushwood, straw, or other things which might be brought down with the stream.

Fig. 73 gives an example of the syphon arrangement as constructed with two cast-iron pipes placed side by side.

Railway works carried out in cities and large towns, whether they take the form of cuttings, embankments, arching, or tunnels, are certain to cause a very considerable disturbance of existing drains, corporation sewers, gas-pipes, water-mains and underground telegraph wires. Some of these underground works may be so peculiar and complicated as to necessitate a slight deviation from the course originally intended for the line. Suitable provision will have to be made for each of the items interfered with by the railway, and the substituted work must be carried out to the satisfaction of the constituted authorities within the municipal boundaries.

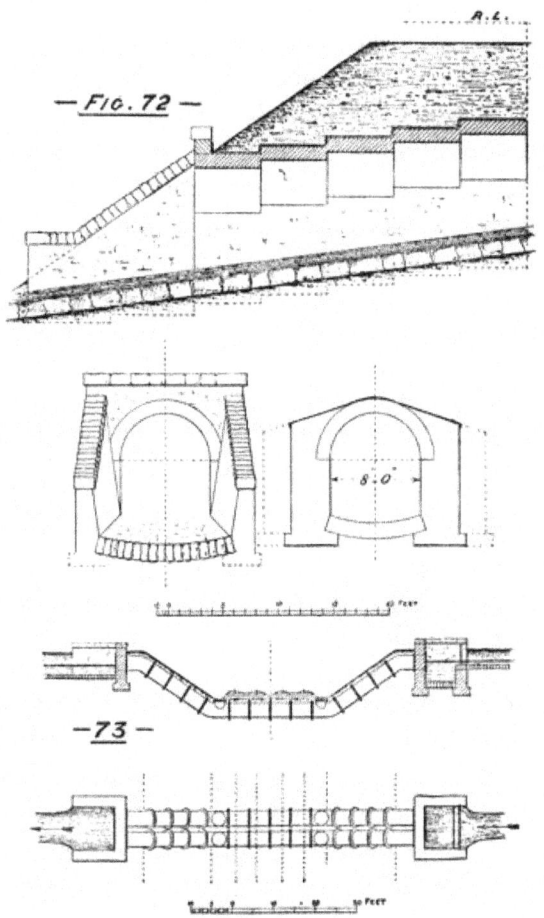

Bridges.—Amongst the many bridges and viaducts which have to be built during the making of a railway those constructed over rivers and waterways are generally the most important The bridging across any navigable river or tidal water can only be effected in compliance with conditions imposed by the authorities controlling the navigation rights. These conditions will place restrictions as to the number and distance apart of the piers, as well as the height from high water level to the under side of the arches or girders. For rivers having a constant traffic of sea-going vessels of large tonnage and lofty masts the authorities will demand great height or headway as well as large spans; and if to this be added a deep water-way and bad foundations, the work to be constructed becomes one of considerable magnitude. The banks of the river must be carefully studied to find the most favourable point for crossing, and in some cases it may be prudent to make a detour of two or three miles. The crossing at a great height involves the construction of the approach lines at a great height also. If the river is in a deep valley with high sloping sides the natural contour of the ground facilitates the formation of the approach lines; but with a river on a low, wide, open plain, inclined approach lines add enormously to the cost of construction, as well as to the cost of permanent working.

If the number of sailing craft passing up and down the river be moderate, and, perhaps, only passing at high water, the authorities may permit a low-level viaduct with an opening bridge.

There are thus the two systems: the high-level viaduct, which allows trains to pass over and vessels to pass under at any and all times, and the low-level viaduct with opening bridge, which, if open for vessels, is closed for trains, or *vice versâ*.

Every crossing of a navigable river will have to be considered and dealt with according to its own individual requirements. An arrangement suitable for the one may not be admissible or prudent for the other. A frequent and important train service might be much interfered with by an opening bridge, and, in a similar manner, an opening bridge might cause much interruption and detention to the navigation of the vessels on the river.

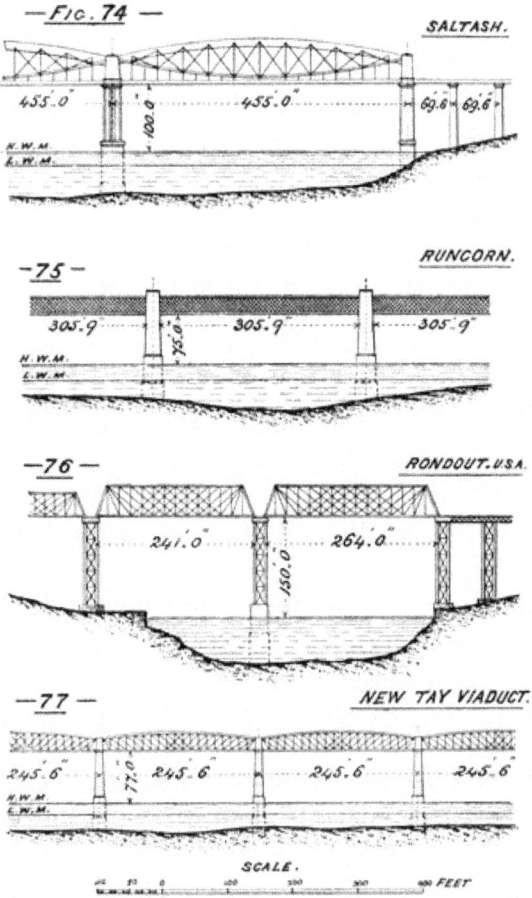

Where a low-level viaduct with opening bridge can be adopted, there will be a very great saving of expenditure; and there are numbers of such viaducts in existence, accommodating a large railway and river traffic without inconvenience. Even with a low-level viaduct the height from water-level to the under side of the girders of the various fixed-spans will generally be sufficient for the passage of barges and small craft, leaving the opening portion to be used by the larger vessels.

The principal openings for these large river viaducts are generally constructed for girders, partly on account of the greater facility of girder work for large spans, and also for the advantage of having one uniform height, or headway, from pier to pier.

For a high-level viaduct across a deep-water river, the cost of the lofty piers forms a very important part of the undertaking. Each pier will require its own cofferdam, caisson, or other appliance for obtaining a suitable foundation. The deeper the water, the more costly the arrangement for foundation; and the higher the pier to rail-level, the greater the amount of material in the construction of the pier. The consideration of these two points will at once show that it is very desirable not to have more of these costly piers than is actually necessary, and in studying out the design it will be a question for calculation how far the spans may be increased so as to dispense with one or more piers.

In every work of this description there is a relative proportion between span and height, which will give the most economical result from a cost point of view; the proportion varying according to the depth of the water and description of ground for foundations. An increase in the span will naturally necessitate an increase in the thickness of the pier; but where a cofferdam, or arrangement for putting in the foundations, must in any case be made, a small addition to its width may not necessarily form a large increase to its cost.

Figs. 74, 75, 76, and 77 are sketches of high-level railway viaducts which have been constructed with great height, or headway, to allow large vessels to pass under at all times without interruption. This description of work is very costly, not only in the deep-water foundations, but also in the heavy scaffolding and appliances requisite for building piers and girders at such an elevation above the ground-level. The hoisting of the material alone forms an important item where such vast number of pieces have to be lifted to a height of 80, 90, or 100 feet.

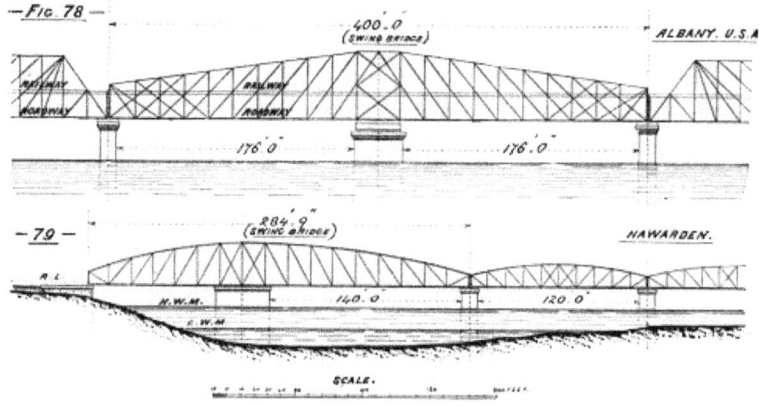

Figs. 78 and 79 are sketches of low-level viaducts constructed with one large opening span, or swing-bridge, for the passage of vessels. The girders and roadway of such opening span are usually constructed as a compact framework, which revolves on a centre placed in the middle of a circular roller path or species of turn-table. The portions of the rotating opening bridge, although not always the same length on each side of the centre-pin, are generally very carefully balanced, to preserve the equilibrium of the entire mass when swinging round for the passage of vessels. To ensure stability in working, and steadiness during heavy gales, a liberal diameter should be given to the roller path of all swing-bridges having large span and great weight.

Lattice, or truss, girders are preferable to plate girders for swing-bridges of considerable opening, as they present less surface area to the action of the wind.

The opening and closing of these bridges is effected by wheel-gearing actuated by hydraulic, manual, or other motive-power. The revolving machinery should be set solid and true, well protected from the weather, and, at the same time, readily accessible for constant inspection, lubrication, or repair.

Figs. 80 to 85 are sketches of various types of railway bridges constructed for smaller openings across narrower rivers, water-ways, or canals. Fig. 80 is an example of what is known as a *bascule* bridge. This particular bridge is made in two halves, meeting in the centre of the span, the tail end of each half being provided with heavy counterweights to assist in opening or tilting up the bridge for the passage of vessels, or lowering it down for railway traffic. Each half of the bridge swings on horizontal axles, and the raising or lowering is effected by means of hand winches or other motive-power, actuating wheel-gearing working into toothed vertical segments attached to the tail end of each half. The same principle has also been applied to bridges having only one leaf to tilt up to clear the passage way.

Railway bridges of this pattern are now very rarely adopted. They have the great drawback that when raised to the vertical position, a very large area is presented to the action of the wind, and this defect might lead to very serious consequences in the case of a bridge situated in an exposed locality. An open-work floor diminishes the wind area, but a very large surface must necessarily remain.

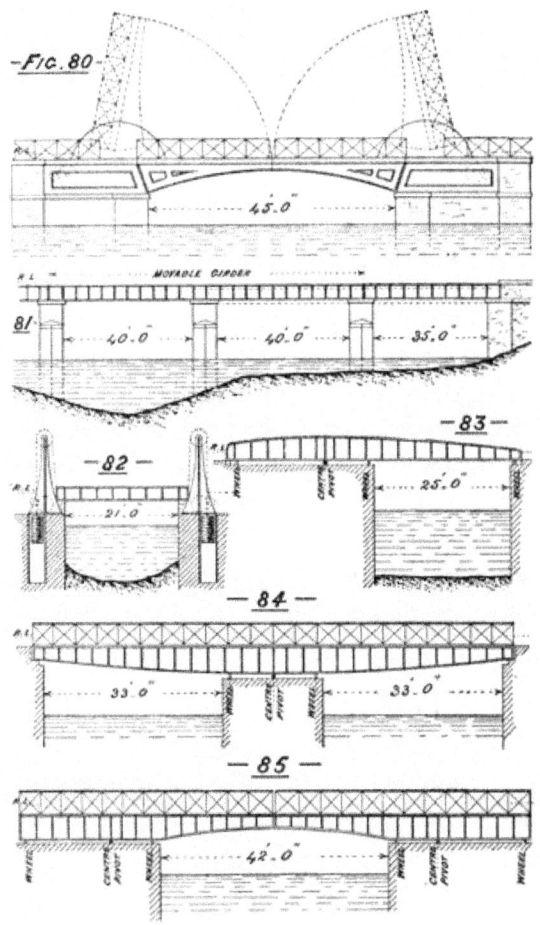

Fig. 81 illustrates what is known as a *traversing bridge*. In this case the width of the opening passage-way and the adjoining span are made the same, and the girders for the two spans are constructed in one continuous length. By means of gearing attached to the fixed portion of the work, the continuous length of girder, with its roadway, is first slightly raised or lowered, and then drawn back on rollers sufficiently far to leave the opening span quite clear for the passage of vessels. A reverse movement of the gearing causes the movable girders and roadway to travel back and return to their original position ready for the train traffic.

Opening bridges are sometimes constructed on this system in cases where the level of the rails is only a few feet above the level of the water, and where there is only one water opening, and that not more than 20 to 30 feet wide. In such bridges the movable portion is rolled back along iron rails, or plates secured to masonry walls, or strong pile-work. This class of bridge is cumbersome, slow to move, and is now but very rarely adopted.

Fig. 82 shows a type of simple *lift* bridge, of which there are but few examples remaining. In this particular bridge the girders and roadway form a solid framework, which rests on the abutments during the passage of the trains. Strong chains, secured to the corners of the framework, pass over large sheaves on the top of the iron standards, and then round drums placed below the level of the rails, and terminate by attachment to heavy counter-weights suspended in iron cylinders. The counter-weights are adjusted to approximately balance the bridge, so that a moderate power applied to the wheel-gearing on the drums is sufficient to raise the roadway to the required height. This class of opening bridge is only suitable for the passage of barges and small craft without masts; and it requires the re-adjustment of the counter-weights when the roadway varies in weight, in consequence of rain or repairs.

Figs. 83, 84, and 85 are sketches of small *swing*-bridges constructed for narrow waterways. Although differing in appearance, they are all practically on the same principle, with centre pin and roller path, and are similar in general arrangement to the large-size-opening swing-bridges shown in Figs. 78 and 79.

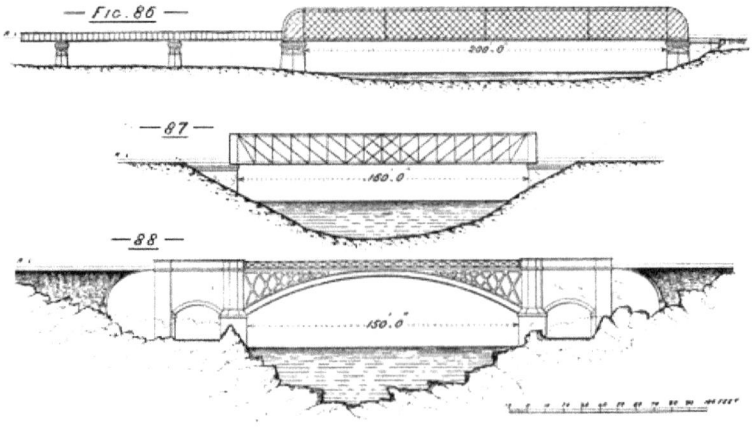

The *swing*-bridge arrangement is so simple in construction, convenient for inspection, and easy to maintain, that where possible it is now generally adopted in preference to any other system. The weights on centre pin and roller path may be distributed as considered most expedient, and by means of suitable appliances the weight may be altogether taken off the centre and rollers when the bridge is closed for the passage of trains.

There are many wide rivers which, although not navigable in the ordinary acceptance of the term, nevertheless require bridges of large spans to provide free waterway for the floating down of rafts of timber. Away in the high ground, in the timber-growing districts, trees are felled, sawn or cut into long poles, logs, or scantlings, and hauled to the banks of the river. The timbers are then formed into large rafts of the most convenient form for floating down to the place of distribution or port for shipment. Even with old experienced floaters, using their long sweeps in the most skilful manner, it is difficult to take anything but a very irregular course down the stream. Under the most favourable circumstances one of these large rafts is an unwieldy, awkward craft to manage; but in a river full of twists and turns, with reaches varying from comparative smooth water to miniature rapids, the current carries the huge mass surging along, and only a clear, unobstructed channel will enable its navigation to be carried out with safety. The presence of a pier in the main waterway might cause destruction to the rafts and loss of life to the men. The vested interests in floating rights are tenaciously guarded, and no new bridge would be sanctioned which would in any way interfere with the waterway or endanger the passage of rafts down the river. Bridges of this description are much less costly than those over deep water— navigable rivers. Excepting the large spans, the rest of the work is comparatively simple. The water is generally shallow, and much reduced in quantity during the summer months. Good foundations can generally be obtained without going to any great depth. The headway may be kept low, or of such height as may best suit the purposes of the railway, and be sufficiently well up out of the way of the floods which may take place from time to time on the river.

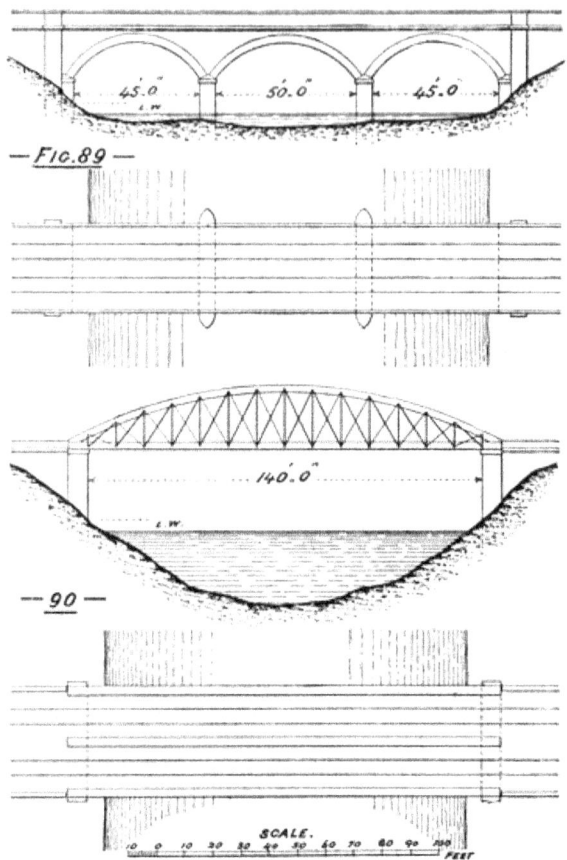

Fig. 86 is a sketch of a bridge constructed over a river much used for rafting purposes. The large span is over the main channel, and the small spans are over a wide gravelly foreshore, which is only covered with water during exceptionally high floods in the autumn or winter. No rafting can be carried on when the river is in flood; the current would be too strong to permit of the raft being kept under control.

Fig. 87 is a sketch of a similar bridge where the river is confined to a regular channel between two sloping banks of strong clay.

Fig. 88 shows a bridge erected over a narrow rocky pass in the river. The channel is hemmed in by the almost perpendicular sides of mountain granite, there are no banks to overflow, the flood waters cannot spread laterally, however much they may increase in depth, and with building-stone at hand in abundance, and foundations formed in the solid rock, the situation is one of the most favourable for a strong permanent bridge. The cast-iron arch of 150-feet span has a graceful appearance, and harmonizes well with the surrounding scenery. A small masonry arch at each end of the bridge provides for communication along the banks of the river.

With rivers which are neither under the control of navigation authorities nor used for rafts of timber, there is much greater freedom for the designing and carrying out of bridges or viaducts suitable for the actual physical conditions of the locality. The headway will be guided only by the height of the railway to be carried across, and by any flood-water levels which may affect the work. The size of the spans will be regulated by the width of the river, the depth of the water, and the nature of the ground into which the piers have to be built. For broad, shallow rivers with good firm river-beds, piers may be built at moderate cost, and comparatively small spans adopted; on the other hand, with a broad deep river it will be better, as previously explained, to reduce the number of piers and increase the span. In the one case, for example, a river 150 feet wide may be crossed with three spans and two piers in the shallow water, as in Fig. 89; in the other it may be more prudent and economical to cross in one span, without any intermediate pier, as shown in Fig. 90.

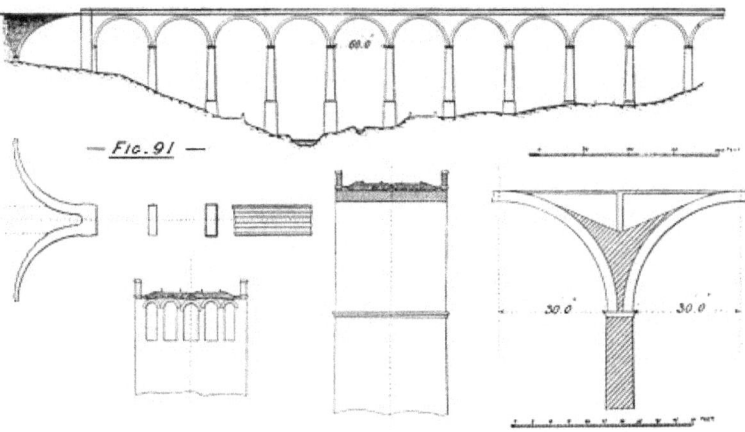

Next in importance to the large bridges and viaducts over rivers are the viaducts which have to be constructed for the crossing of deep inland valleys. The occurrence of one of these deep valleys between long lengths of average table-land renders necessary either a series of cuttings and falling gradients to get down to a low level, or the erection of high-level works to continue onward the rail-level at the height already attained. A decision to adopt the latter course brings forward the consideration as to the method of carrying out the work. To form a high embankment across such a valley would entail an enormous expenditure for earthwork, and several openings, or bridges, would have to be made in the embankment for streams, rivers, and roadways. Instead, therefore, of making this part of the line entirely of embankment, it is usual to carry the earthwork forward until the height is about 25 or 30 feet, and to form the remainder of the opening of arching, as shown in Fig. 91.

This arrangement is not only less costly than an embankment of such height, but has also the great advantage that any or all of the arches are available for the passage of streams, rivers, roads, and accommodation works.

The character of the work to be carried out in the construction of bridges or viaducts over rivers or valleys must greatly depend upon the description of materials at command. Where good building-stone is plentiful, and the price of labour moderate, works of masonry should be adopted as far as practicable. Brickwork is an excellent substitute for masonry, provided that specially selected bricks are used for all facework, or parts exposed to the weather. For water-washed piers and abutments, the lower portion should be faced with good hard stone.

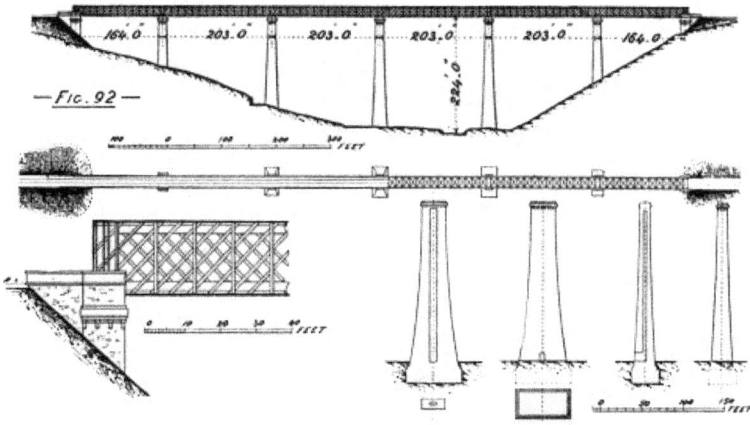

—FIG. 92—

Bridges and viaducts consisting of arches of masonry or brickwork form the most substantial and permanent works of construction for railway purposes; once properly built, the expenditure on future maintenance or repairs is merely nominal. For viaducts the span of the arching must be regulated by the height of the viaduct. The greater the height the larger the span. In one case 30-feet spans may be suitable, whereas in another it may be more economical to introduce spans of 60 feet or more, and so reduce the number of lofty piers. From a cost point of view there is, however, a limit to the span of arching, and, except for special cases, where expenditure is of secondary importance, large spans are very rarely adopted. Arches of large spans, no doubt, have been built both in masonry and brickwork, and have been a complete success in every way except expense. Unfortunately, the quantity and weight of materials in arching, and the corresponding cost, increase very rapidly as the span increases, and for openings of more than 60 or 70 feet girder-work becomes much cheaper than arching.

Figs. 92 and 93 are examples of viaducts having piers of masonry, with girders to carry the roadway. In the one case the roadway is carried on the bottom flange of the girders, and in the other on the top. The latter arrangement affords greater facility for securely bracing the girders together, while for the former it is claimed that the girders form a massive parapet, which would serve as a protection in the event of an engine or vehicles leaving the rails.

In the early days of railways, many large viaducts were constructed having masonry piers, and timber trusses to carry the roadway. Much ingenuity was displayed in designing the trusses, and in the introduction of cast-iron joint-shoes and wrought-iron bracings. Many of these wooden superstructures served well for several years, but they were always exposed to the imminent risk of destruction from fire, and however carefully the logs may have been selected, the decay of the timber was only a question of time. The deterioration of one piece was equivalent to the weakening of the entire truss, and the renewal of any part was both difficult and costly. The shrinkage of the timber, and the working at the joints, caused the trusses to deflect considerably under a passing load, and although the actual strength of the structure may not have been much impaired, the creaking and depression had anything but a reassuring effect. Timber superstructures for anything but small spans are rarely adopted now, except for temporary works, or on lines abroad, where the transport on girder-work would be very costly, and where good timber is very cheap and abundant. Even in the latter case the wooden superstructure is generally looked upon as a temporary expedient, to be replaced at no very remote date with iron or steel girders, when the materials can be conveyed over the entire completed line.

Figs. 94, 95, and 96 are sketches of three types of timber trusses as constructed in viaducts of several spans.

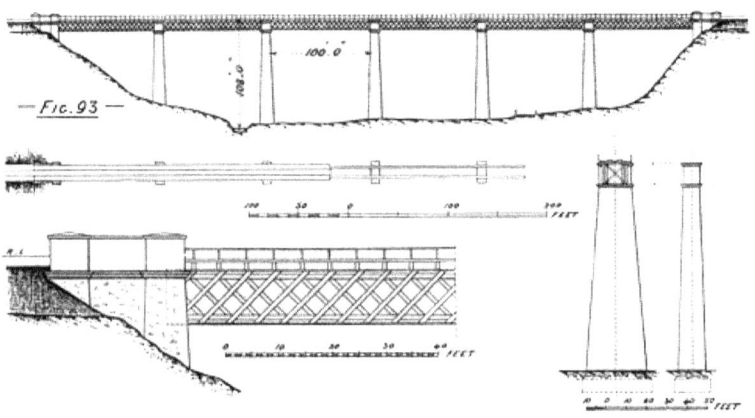

There are many localities, especially abroad, where suitable stone is most difficult to obtain, and very expensive to work and convey. In such cases it is compulsory to use as little of it as possible, and to resort to iron or steel both for the girders and a large portion of the piers. The piers may be made of cast-iron, wrought-iron, or steel, of suitable form and arrangement to ensure strength and stability. Not only must the piers be strong enough to carry the weight that may be brought upon them vertically, but they must have sufficient width of base to ensure lateral steadiness. The design should admit of facility of erection, with a minimum of scaffolding, and the pieces should be of convenient length and weight for transport. The lower length of river piers, or portion liable to be in contact with flood-water, should be of solid masonry, to resist the action of the water, or of any *débris* brought down by the current. More than one fine viaduct has been swept away for want of due attention to the latter precaution.

Fig. 97 illustrates a type of pier composed of cast-iron columns, well braced and stayed with wrought-iron. The ends of the columns and all contact surfaces should be properly turned and faced by machinery to ensure true and perfect joints, and the socketed ends should be turned and bored to fit closely. The latter is important, and if not carefully carried out, a slight sliding movement of the flanges may take place, and throw undue strain on the bolts.

Fig. 98 shows a very similar pier, constructed entirely of wrought-iron or steel.

Each of the above-described piers has a liberal amount of taper or batter, both in the front and transverse elevation.

The size and number of the columns, and the dimensions of the braces or stays, will depend upon the height of the pier and the weights and strains to be sustained.

Many important and lofty viaducts have been erected on this principle of iron piers springing from masonry foundations, more particularly across deep rugged ravines abroad, where iron piers offered the only practical, substantial means of dealing with what appeared otherwise an impossibility.

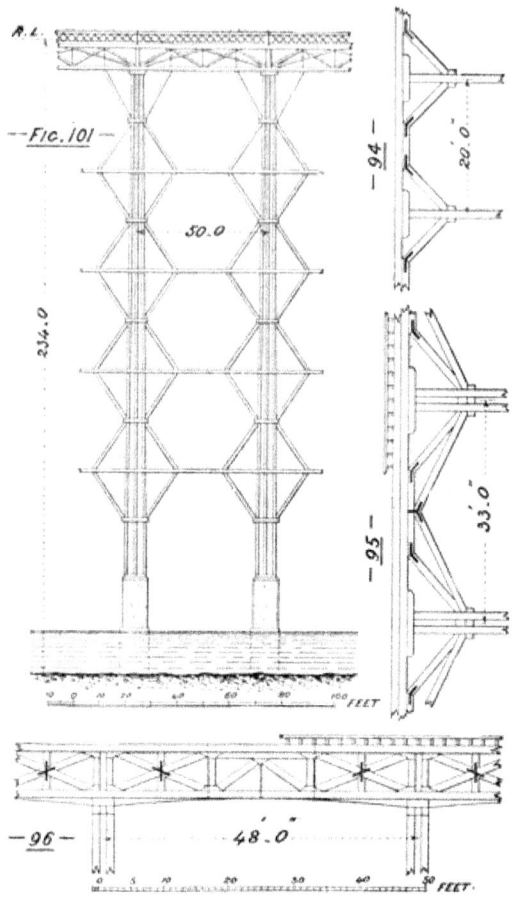

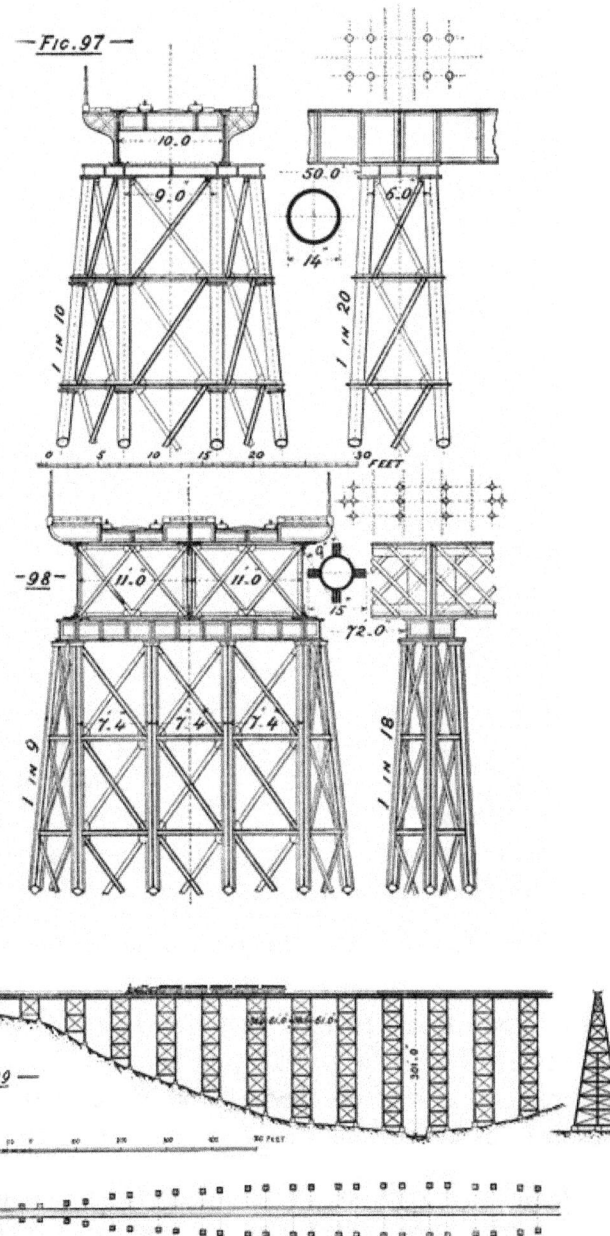

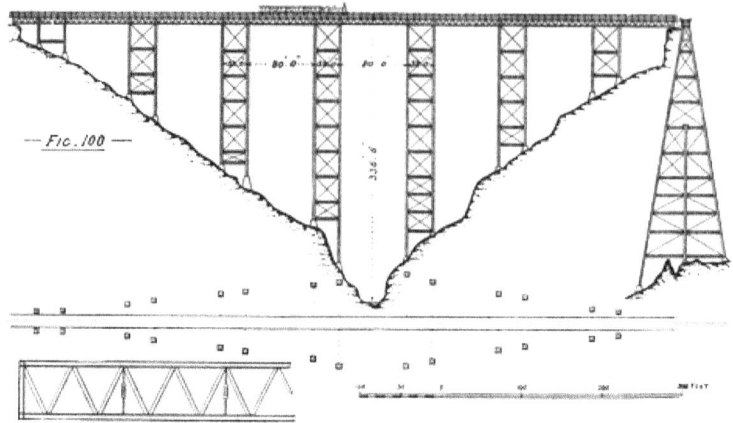

Fig. 99 is a sketch of the Kinsua Viaduct on the Erie Railway, one of the highest railway viaducts in the United States. In the transverse elevation the piers have a large amount of taper; but in the front elevation they are vertical, and of width to correspond to one of the small spans of the main girder. This arrangement of long and wide base gives great stability to the pier. The spans of the girders, which are of the ordinary lattice type, are not large, being 61 feet for the clear spans, and 38 feet 6 inches for those over the piers. The principal interest is in the great height and simplicity of the piers. The rail-level over the top of the pier is 301 feet above the level of the water in the Kinsua stream. The width of this pier on the top is 10 feet (for single line), and the width at the bottom 103 feet.

Fig. 100 is a sketch of the Loa Viaduct on the Antofagasta Railway, Bolivia, stated to be the highest railway viaduct in the world. The arrangement of spans and piers is very similar to the Kinsua Viaduct. The main spans are 80 feet, and the pier spans 32 feet. The width of the pier on the top is 10 feet 6 inches (for single line), and the width at the bottom of the highest pier is 106 feet 8 inches.

In contrasting these light iron piers with what would have been required if constructed of masonry, an idea may be formed of the enormous amount of material, labour, and time, which would have been expended to erect the work in stone.

Before the principle of lofty iron piers had been thoroughly developed, many high piers had been built of timber both at home and abroad. More particularly was this the case in the United States of America, where the presence of magnificent timber close to hand offered special inducements for the use of wood. Like a mammoth scaffolding, each pier was constructed with a most liberal supply of material, judiciously selected and carefully put together, but the danger of destruction by fire was ever present from the beginning. Probably more timber piers and bridges have been destroyed by fire than have been removed on account of natural decay.

One of the most notable of these timber-pier constructions was that of the Old Portage Viaduct, on the Erie Railway, U.S.A. Fig. 101 is a sketch of one or two of the piers. This viaduct was more than 800 feet long, and 234 feet high from the bed of the river to the rail-level. The spans were 50 feet each. Masonry piers were carried up to about 25 feet above the ordinary water-level of the river, and upon these the timber superstructure was erected. Each timber pier consisted of three complete sets of framework, securely connected together, and also well stayed and braced to the adjoining piers. This viaduct was destroyed by fire in 1875, and was reconstructed with piers and girders of iron.

Railway bridges over or under public roads of primary or secondary importance must be constructed to the widths and heights prescribed for such works in the fixed regulations of the country in which they have to be built. As a rule, these road-bridges are simple and inexpensive in character, except in towns, or in cases where the line crosses the roads very obliquely, or where the road is situated at the top of a deep cutting, or bottom of a high embankment. Away from towns and out in the open country, permission is generally obtained to divert the roads to a moderate extent, so as to obtain a more favourable angle and height for the bridge; but in towns, where the roads become streets, sometimes of great width, with houses and shops on each side, little or no diversion can be allowed.

A railway passing through a portion of a densely populated town must deal with the streets as they exist, as any great alteration in their course or continuity would involve a large destruction of property. With careful laying out it is possible to obtain favourable crossings for many of the streets, but a number of others must be crossed obliquely, and these oblique crossings very frequently result in a span twice the width, or even more, of what would be necessary to cross the street on the square. Bridge-work in towns is more costly than in the country, as a higher class of work is demanded, more finish or dressed work in the masonry or brickwork, and more ornamentation in the screens and parapets in connection with the iron girder-work. The work itself has to be carried on in a confined locality, with limited space for materials and appliances, and where the thoroughfare must be kept open.

Where the height is sufficient, and suitable materials readily obtained, it is preferable to adopt an arch bridge, as being of a much more permanent character than girders.

Fig. 102 is an example of an ordinary over-line arch bridge to carry a public road over a double line of railway in a cutting of moderate depth.

Fig. 103 shows a somewhat similar over-line arch bridge, but its height from rail to road-level being greater, side arches are introduced in preference to long heavy wing walls.

Fig. 104 shows an over-line arch bridge in a rock cutting. In this case, by increasing the span and forming the springing bed in the solid rock, the masonry of abutments and wing walls may be reduced to a minimum.

Fig. 105 is a sketch of an ordinary under-line arch bridge to carry a railway over a public road in an embankment of moderate height.

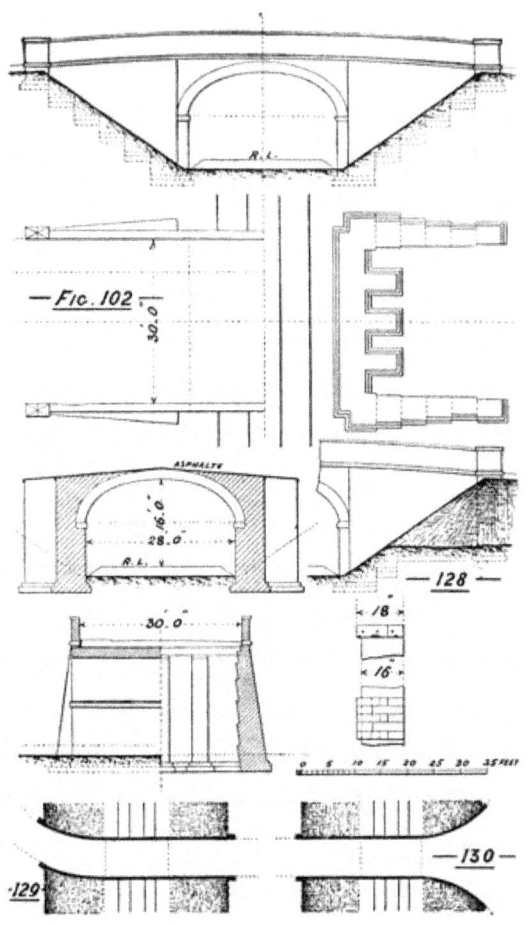

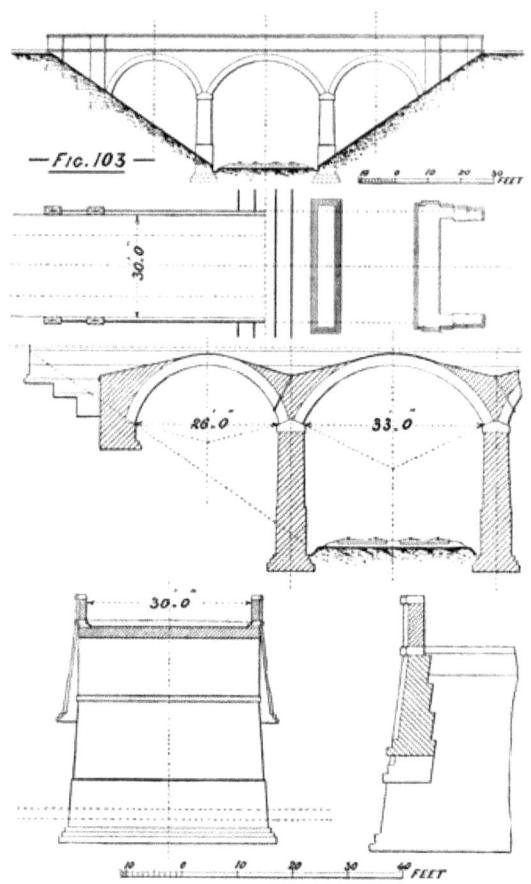

Fig. 103

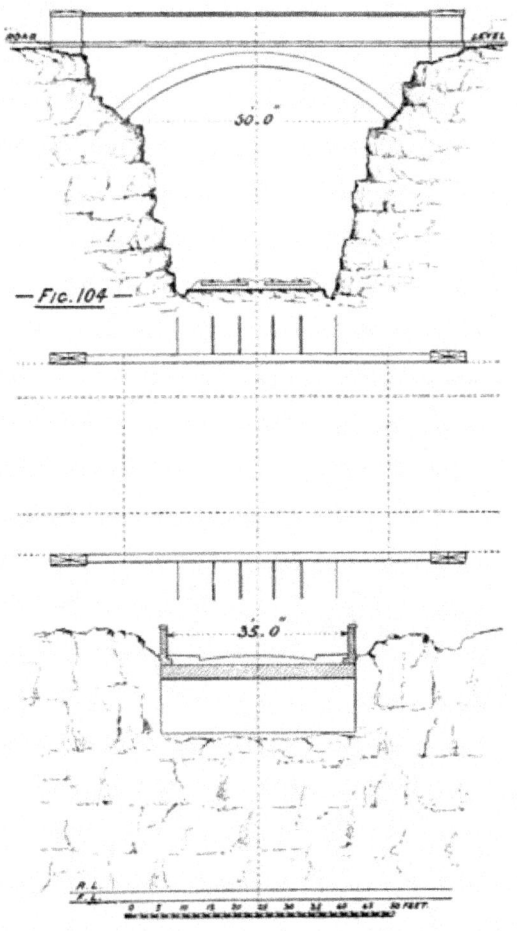

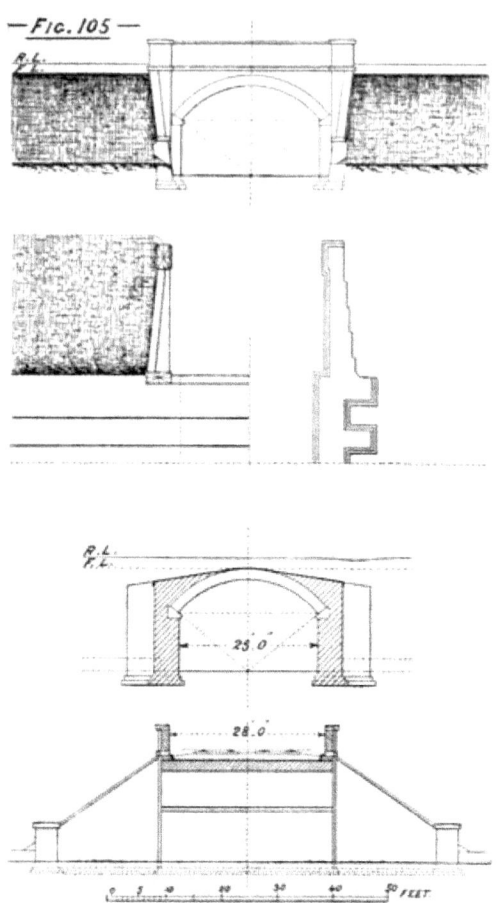

FIG. 105

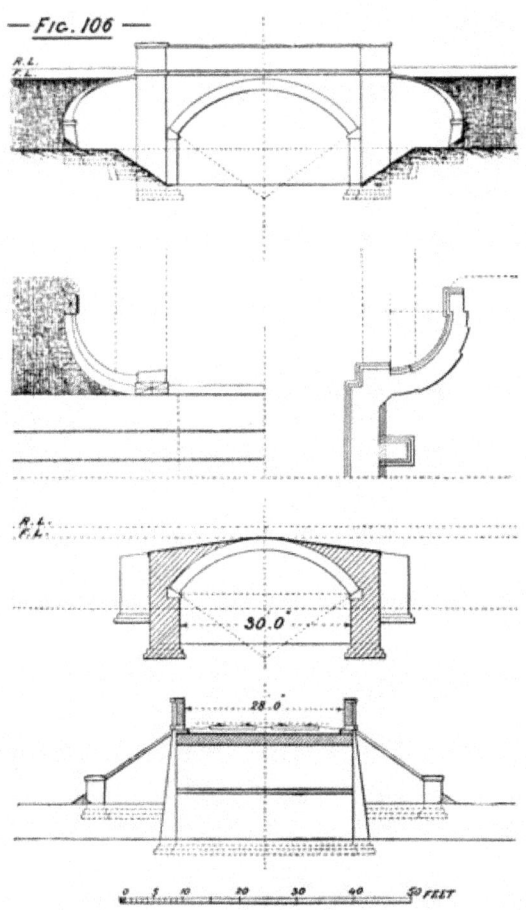

Fig. 106

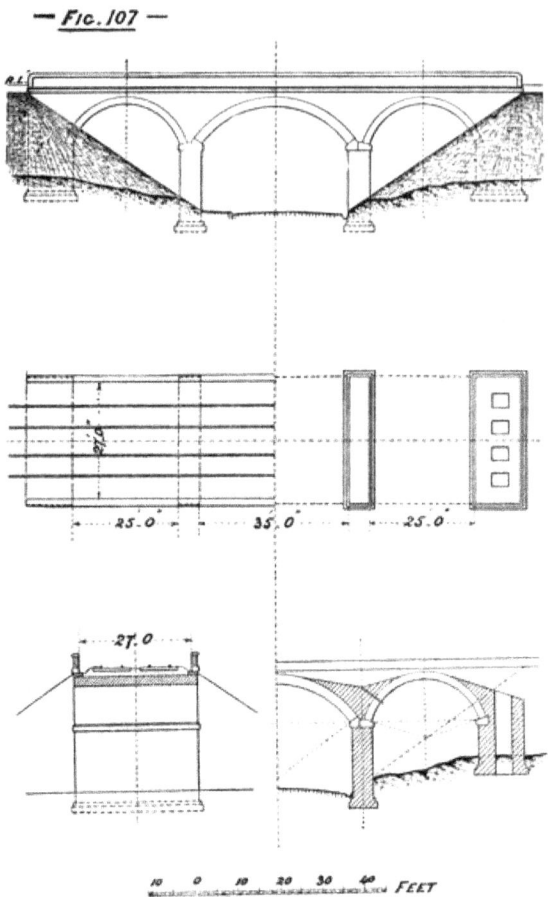

Fig. 106 shows a similar under-line bridge, but with curved instead of straight wing walls.

Fig. 107 is an example of an under-line arch bridge in a rather high embankment, and where side arches have been adopted instead of long wing walls.

The above six types are equally applicable for private roads crossing the railway, but, as previously mentioned, a lesser width and headway will be accepted for under-line bridges for private or occupation roads, than for public roads. For the over-line bridges, however, the width and headway will be regulated by the number of lines and standard height of the railway.

When these arch bridges have to be built on the skew to suit an oblique crossing of the road, extra care will be necessary in setting out the work, and marking on the centering the spiral courses of the arching.

Arch bridges may be built of masonwork or brickwork, or a combination of the two. If the available quarries do not yield good flat bedded stones readily worked, it is better, where possible, to use strong hard bricks for the arching, and utilize the stone for the remainder of the work.

Although arching undoubtedly forms the most durable type of bridgework, numbers of cases occur where the available height or space between rail-level and road-level is too small, or the cost of masonry and brickwork too great, to admit of anything but girder-work. Detailed sketches of some of the many forms of girder bridges are given in Figs. 132 to 153, illustrating various systems of roadways and parapets. In some instances the main girders are made sufficiently deep to serve as parapets, while in others a shallower girder has been adopted, on top of which has been placed a light cast-iron parapet composed either of close plate-work or of ornamental open railings. The open ironwork parapet has a good appearance, but as a screen is not so efficient as the close cast-iron plates.

In addition to the bridges required for the regular public roads, it is usually necessary to construct a certain number of occupation or private road bridges over and under the line to accommodate portions of estates and large properties intersected or severed by the railway, and which would be inadequately provided for by ordinary gate crossings on the level. The position and description of these occupation bridges is generally matter of private arrangement. The bridges will be somewhat similar in character to the public road bridges, but of much less width for the roadway. Those over the railway must have the standard span and height adopted as a minimum for the other over-line bridges, and those under the railway must have the full width on the top for the lines of rails, but will have less width between the abutments for the roadway.

Foundations.—So much depends upon the soundness and security of the foundations of any bridge, viaduct, or large building, that it would be almost impossible to devote too much care to the selection and treatment. Unless the foundation be firm, the entire structure will be exposed to the risk of failure, either in subsidence of masonry, giving way of arches, or depression of girders. A small matter overlooked during the construction of this part of the work will be most difficult to correct or adjust afterwards.

The insistent weight of all structures built of masonry or brickwork will cause the mass to settle to a certain extent, according as the joints of mortar or cement become compressed by the number of superincumbent courses. In a similar manner the gravel and clay of a foundation will compress more or less according to its compactness and the weight of the structure. No inconvenience will, however, arise if the settlement or compression be uniform throughout the entire area.

In ordinary average, dry, solid ground, a good foundation can usually be obtained at a moderate depth. The removal of a few feet of the surface layers will generally lead to a good hard stratum of natural material sufficiently firm to carry the abutments and piers of railway bridges and viaducts. Two or more footings are usually adopted so as to distribute the weight over an increased area, as shown in Fig. 108.

Where the weight to be carried is considerable, it is better to increase the number of the footings, and give them a smaller projection, as in Fig. 109, rather than have a lesser number and greater projection, as in Fig. 108. There is greater liability of fracture of the material in the latter than in the former.

Care must be taken to distinguish between made ground and natural ground. Hollows which have been filled in must not be relied upon to sustain heavy weights; the material may have been consolidating for years, but it is safer to cut through it and found upon the natural stratum beneath.

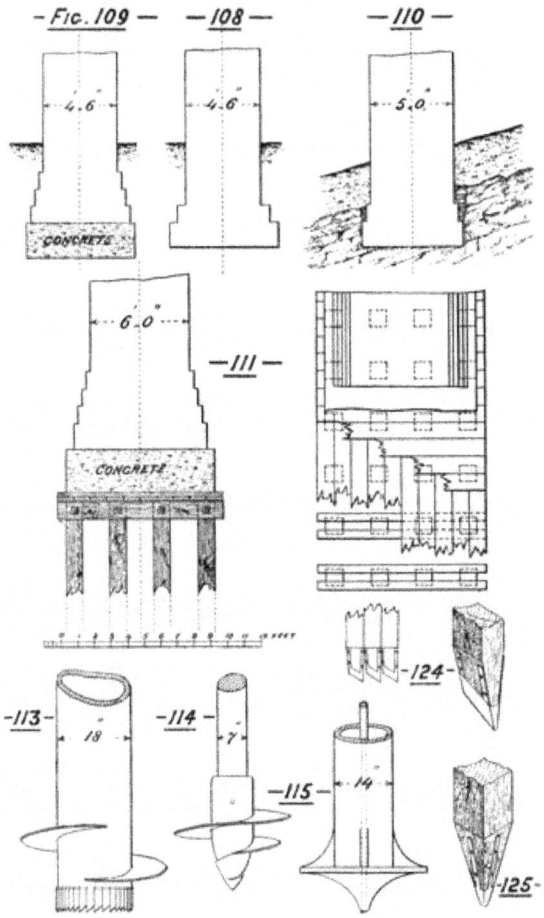

Soils of a clayey nature must be dealt with very cautiously. If the ground be fairly level, and the material firm, a solid foundation may be obtained, but the excavated portion should be covered up as quickly as possible to prevent any decomposing action taking place upon exposure to the open air. The expansive nature of some clays must be carefully kept in view, so as to guard against any disturbance in the finished foundation. There are some descriptions of shale which when first opened out appear to have the solidity of hard rock, and yet, after a few days' exposure to the atmosphere, are changed to the consistency of soft mud.

Sand, being composed of such small particles, is almost incompressible, and makes an excellent foundation so long as it can be retained in its position. Little or no settlement will take place if the sand remains undisturbed, but so soon as it comes under the influence of running springs, or underground drainage, the fine particles of the sand will be gradually but surely carried away with the water, and the entire foundation be undermined. The opening out of a neighbouring excavation, or the carrying out of some low-level drainage, would endanger a construction which otherwise would be solid and permanent.

In many cases of soft ground, more particularly abroad, sand piles have been adopted and have given very good results. The system is carried out by first driving a large wooden pile down through the soft material into the more solid stratum below. The timber pile is then carefully withdrawn and the cavity filled with clean sand. The number and distance apart of these sand piles will depend upon the nature of the ground and description and weight of structure to be carried.

Clean, compact gravel is one of the best materials to build upon, being almost incompressible and quite unaffected by exposure to the atmosphere. It is easily excavated and levelled off to the surface required.

A foundation of rock may be considered in the abstract as the most solid base to be obtained, but it must be treated judiciously, and a proper surface secured. The outer portion of many descriptions of rock consists of blocks or layers of stone partially or entirely separated from the main bed, and these, lying in a loose condition, are deceptive and treacherous as a foundation base. The exposed rock should be carefully examined, and all detached or outlying pieces or layers removed before placing any foundation course. Special care must be paid to all shelving rock, and a level seating cut into it for the entire width of the foundation, as shown in Fig. 110.

A thick bed of concrete, as in Fig. 109, makes an excellent foundation course. When firmly set it becomes one solid massive base from end to end, and prevents the yielding or dropping of masonry at any intermediate points.

There are many places in soft, wet ground where instead of attempting to excavate all the soft material down to a harder stratum, it is better to adopt timber pile foundations, as shown in Fig. 111. The size of the piles and their distance from centre to centre must be regulated by the description of material into which they have to be driven and the weight they have to sustain. Double waling pieces should be properly checked and bolted on to the heads of the piles, and trimmed or levelled off to receive a double floor of thick planks. The spaces round the heads of piles and walings should be filled in and levelled up to under side of flooring, with cement concrete.

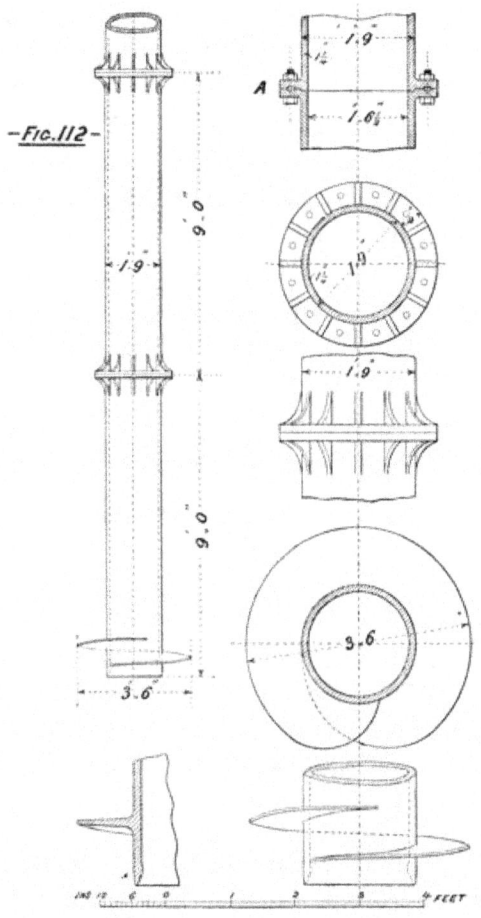

For bridges of moderate span, over soft ground or over shallow fresh water, strong cast-iron screw piles can be adopted with great advantage. Fig. 112 shows a very usual form of screw pile, made with an external screw at the lower end and with a sharp cutting edge to facilitate penetration into the ground. The upper portions are made in suitable lengths, and all to one pattern and template, for convenience in carrying out the work. The screwing into the ground is generally effected by means of a capstan or cross-head fixed to the top of the first working length of pile, and which is pulled or turned round by ropes worked from stationary windlasses. In some cases long bars or levers are attached in radiating positions to the capstan-head, and a number of men are employed to walk round and round, pushing the levers, and in this way screwing the pile into the ground. As the pile goes down the capstan-head has to be removed, and additional lengths bolted on, until the pile enters a solid stratum, or is considered deep enough for the duty it has to perform. The last or top length has generally to be cast to a special length to bring the work up to the exact height to receive the girders. The core of excavated material passes up into the interior of the pile, and in some cases becomes so compressed or tight as to require the use of an internal augur to remove a portion of it to enable the screwing to proceed. The pile shown in Fig. 112 is one of a number which were successfully screwed into the ground to depths varying from 42 to 48 feet. A toothed or serrated edge, as in Fig. 113, is sometimes given to the lower edge for screw piles which have to cut their way through a hard stratum.

All bolting flanges should be accurately turned and fitted to ensure close, parallel surfaces when bolted together.

The joint shown at **A**, Fig. 112, is one the writer has used to a large extent for the bolting flanges of cast-iron screw piles and cylinders. It is very simple in form, readily coated with white lead to ensure a water-tight joint, and as the upper length is practically recessed, or let into the lower length, the exact continuity of the different castings is secured.

Solid screw piles of wrought-iron or steel, similar to Fig. 114, are used for some descriptions of work. These are generally made in long lengths, in sizes varying from 4 to 8 inches in diameter, and with screw blades of wrought-iron or cast-iron fixed in the most secure manner to resist the strain produced when screwing into the ground. The couplings for these solid piles must be very carefully made, all contact surfaces truly faced and fitted, bolts turned, and bolt-holes drilled.

Fig. 115 is a sketch of a hollow cylindrical water-jet pile, which has been used successfully in cases of light sand. The lower end of the pile is made externally in the form of a solid disc, terminating in a conical point, having an aperture in the centre to correspond to the water-jet. To the top of the pile is secured a tight-fitting cover through which a tube passes from a force pump. Water at high pressure is pumped into the tube, and as it forces its way out through the conical point the sand is stirred up and loosened, and thus allows the pile to descend. When the pile has been lowered to a sufficient depth the pumps and tube are removed, and the sand settles down into its former compact condition.

Great care must be used with the first two or three lengths of any screw pile to ensure the pile taking a correct or true vertical position. Each series of screw piles should be properly braced together to obtain stability under moving loads.

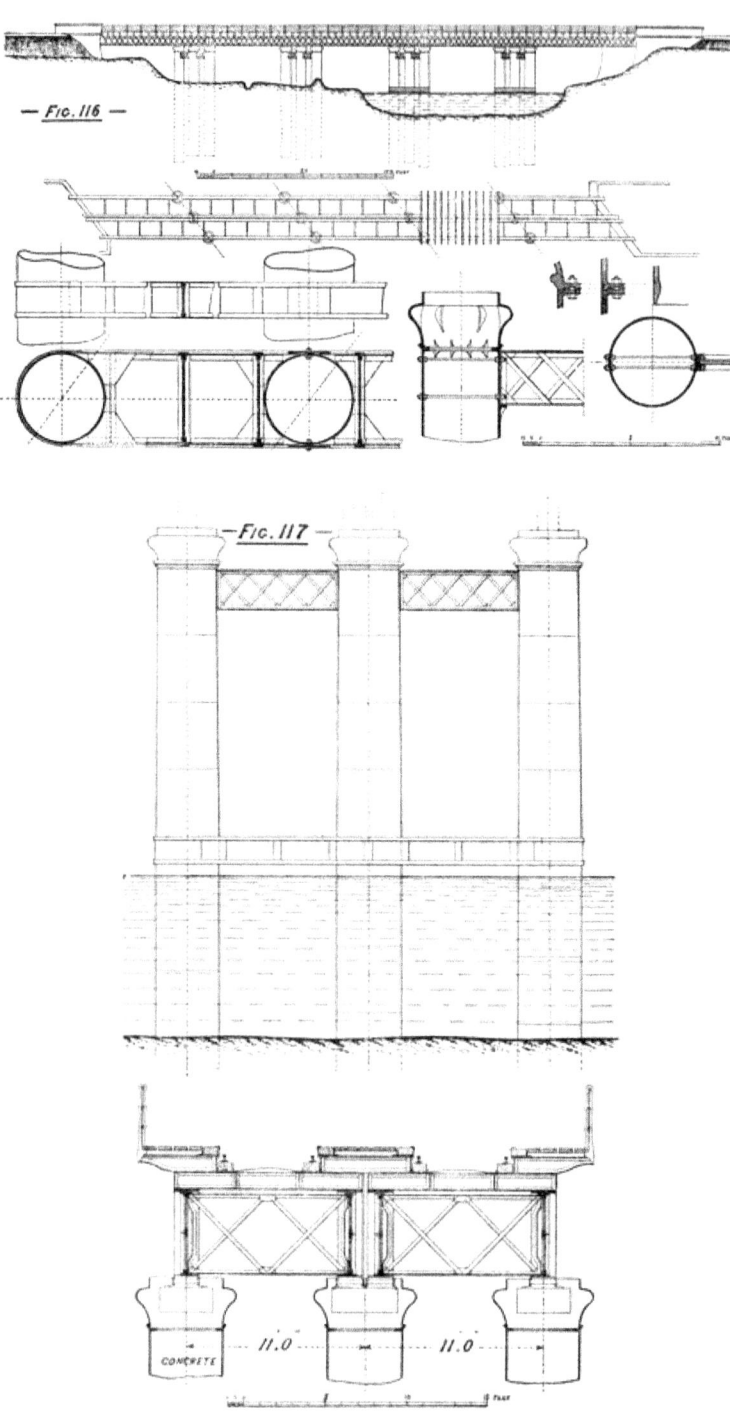

Hollow cylinders of cast-iron, wrought-iron, or steel form most efficient foundations or piers for large bridges over soft ground or fresh water of considerable depth. Made open at the bottom, and constructed of complete rings, or, if of large diameter, of rings built up in segments and securely attached together with water-tight joints, the cylinder is placed in its proper position on the ground or lowered into the water preparatory to sinking. The lower length is made with a sharp cutting edge to facilitate penetration. By excavating and removing the material round the cutting edge and base inside the lower length, the cylinder descends gradually either from its own weight or by assisted weights, and length after length is added until it is sunk to the depth required. The excavated material is filled into buckets and hoisted to the surface by a winch fixed on the top length. When sinking in water the working top of the cylinder is always kept at a suitable height above the water for convenience in removal of the earth or clay from the interior to barges or gangways alongside.

Some strata are more favourable for cylinder sinking than others. Material of a strong clayey nature admits but a small amount of water into the excavation, and a moderate-sized pump will keep the working fairly dry until considerable depth has been reached. Some other materials are so open that the water cannot be kept down with ordinary pumps, and the cylinders can then only be lowered by the pneumatic process. This process has been carried out in two methods, one of them on the *vacuum* principle, and the other by air pressure, or, as it is termed, the *plenum* system. With the former method the cylinder is placed in position, and an air-tight cap, through which a pipe passes, is secured on the top. Powerful air-pumps are then set to work, and the partial vacuum thus created in the interior causes the material round the cutting edge and base to be loosened and drawn into the cylinder, the cylinder at the same time going down or sinking by its own weight, or assisted, if necessary, by added weights. The cap is then taken off, and the material removed from the interior, the operation of exhausting and emptying the interior being repeated until the cylinder is sunk to its proper depth. This method has been found to work well in strata which contained a large proportion of clay to assist in excluding the air and water, but was not nearly so successful when applied to material containing stones and large boulders.

The *plenum* process is based on the principle of the diving-bell, the water being prevented from entering at the bottom by keeping the cylinder full of compressed air. An air-chamber, or *air-lock*, with perfectly air-tight joints, is securely fixed to the top or upper working length of the cylinder, and no access can be obtained to the interior of the cylinder without passing through this air-lock, which has one lower door or valve opening into the cylinder, and an upper door opening out into the open air. Temporary inside staging is formed by putting planks across from flange to flange, and placing short ladders on these landings for the use of workmen descending or ascending. The excavated material is hoisted by a winch, generally placed on the landing just under the air-lock. The air-pump is placed in some convenient position outside, near at hand, the pressure-pipe passing through the air-lock into the interior of the cylinder. Air is forced into the cylinder to a pressure sufficient to drive out and keep out the water from the interior, and allow the workmen free access for excavating the material round the cutting edge and base of cylinder. The amount of pressure required will depend upon the depth of the working below the level of the water alongside. Men accustomed to the process can work without much inconvenience under a pressure of 20 to 22 pounds per square inch, equal to a depth of 45 to 50 feet; but when the pressure exceeds 25 pounds, the duty becomes very trying, and is attended with considerable risk. Instances are recorded of men working at depths of 105 and 110 feet, necessitating a pressure of over 45 pounds per square inch; but it is very questionable whether the men exposed to such a severe ordeal were not permanently affected, if some of them did not actually succumb.

It will sometimes occur that, after sinking through soft porous strata to a considerable depth, a layer of clayey material is penetrated sufficiently retentive to keep out the water and permit of the removal of the air-lock and the completion of the sinking as an open-top cylinder.

When working on the *plenum* system everything must pass through the air-lock, both materials and men. The excavated material is hoisted up to the level of the air-lock, the upper and lower doors of which must be closed, and the pressure inside the air-lock brought to the same as that inside the cylinder by means of a regulating valve. The lower door is then opened to admit the excavated material, and then closed again to cut off all communication with the interior of the cylinder. The upper door is then opened, and the material hoisted out into the open air. The same process has to be adopted for the egress of the workmen, and the reverse arrangement for the ingress of men and materials. The shape and dimensions of the air-lock may be varied according to circumstances, but the principle will remain the same.

When the cylinder has been lowered to what is considered a sufficient depth, it is usually loaded with a certain amount of dead weight in the shape of old iron or other convenient material, and allowed to remain loaded for some days to ascertain if it will sink any further. Should this test be found satisfactory, the dead weight is removed, and the interior of the cylinder pumped dry and carefully filled with good cement concrete.

Cylinders for foundations are generally made circular in section, that form being the most convenient for turning and facing the flange-joints. They can, however, be made oval in section, or of any section that may be found most suitable for the work required. Figs. 116 and 117 give the particulars of a double-line railway bridge carried on cylinder piers across a river. The detail sketches explain the form of cutting edge, flange joint, and method of bracing. This bridge is one that was reconstructed and widened from a single-line to a double-line bridge. Traffic was carried over on one line while the second line was being erected, hence the reason why one strong central girder was not adopted.

Cylinders of 7 feet diameter and upwards are sometimes filled with concrete in the lower portion, on which is built either a circular lining or a solid mass of masonry or brickwork up to the level of the girder-blocks. In some cases the cylinders proper, together with their concrete filling, terminate a little above the water-level, and upon these foundations are erected strong cast-iron columns, plain or ornamented in design, to carry the girders and roadway. The cylinder itself is generally considered merely as a casing or medium for obtaining a foundation, the weight of the superstructure being carried on the internal filling or lining.

Caissons constructed of plates of wrought-iron or steel are much used for the foundations of large piers in deep water. Practically they may be considered as cylinders on a large scale, with the difference that whereas cylinders are generally continued up to the under side of the girders of the superstructure, caissons are only carried up to a short distance above the water-level. A caisson forms a strong water-tight iron cofferdam, from which the water can be excluded, and a masonry or brickwork pier constructed inside. It may be made all in one piece to correspond to the form of the pier, or in separate pieces to form one whole, each being sunk independent of the other, and connected together afterwards. Being built up of plates cut to the proper size and shape, it is a very simple matter to rivet on additional tiers of plates as the caisson is lowered deeper and deeper into the bed of the river. The lower length is made with a cutting edge to penetrate the ground; the exterior is made without any projection larger than the rivet heads, and the interior is strengthened with **T**-irons or double **L**-irons at the joints, and strong cross-bracing to resist the pressure of the water. About 7 or 8 feet above the cutting edge a strongly framed iron floor is riveted to the vertical sides, and strengthened by plate-iron under-brackets placed at short distances. The excavators work in the space below the floor, and the excavated material is passed up through openings formed in the floor at convenient points to suit the working. The methods of lowering a caisson are the same as for lowering a cylinder. If the pneumatic system has to be adopted, then two or more air-tight tubes of liberal dimensions (say 5 to 8 feet diameter), according to the size of the caisson, must be attached to the floor, and on the top of each of these tubes air-locks must be secured for the removal of men and materials. The masonry or brickwork of the pier is built upon the iron floor, and a portion of this building work is usually carried on during the sinking of the caisson to obtain weight to assist in the lowering. When down to the proper depth, the space below the floor is properly cleared of *débris* and water, and then carefully filled in with cement concrete.

Some caissons are made with vertical sides throughout their entire height; others have an outward taper for 15 or 20 feet on the lower end. The former are not only simpler in construction, but are more easily kept in a vertical position during the sinking. Caissons are usually put together in some convenient place near the edge of the water, and then conveyed on pontoons to the sites of the piers. Great care is required in lowering them into position in the bed of the river, and guide-piles, guy-chains, and other appliances are frequently necessary to keep them vertical during the sinking.

The form, dimensions, thickness of plates, cross-bracing, and general arrangement will depend upon the size and depth of the pier to be constructed. Caissons for heavy work on difficult or treacherous ground require great care, not only in their construction, but also in placing them in exact position, and in sinking them correctly to their proper depth. A tilted caisson is a most difficult subject to handle, and entails heavy expenditure to restore it to a true vertical position. By making careful borings, the engineer can ascertain very closely the depth to which the caisson will have to be lowered to obtain a good firm foundation. With this information the caisson can be so constructed that the upper portion, termed the temporary caisson, commencing a few feet above the bed of the river, can be detached, and removed at the completion of the work from the lower or permanent portion sunk below the ground line.

Fig. 118 gives sketches of a wrought-iron plate-caisson applied to a deep-water river pier, and lowered to its full depth by the pneumatic process; dotted lines show the air-tubes through which the excavated material is hoisted and emptied into barges alongside.

Many large and important pier foundations have been constructed on the system of brick cylinders or wells, particularly in India, where the foundations for large river viaducts have to be carried down to great depths through thick deposits of soft material. These wells are built upon V-shaped curbs to facilitate the penetration when sinking. Fig. 119 is a section of a well with a wrought-iron curb, and Fig. 120 is a similar well with a wooden curb. The wrought-iron curb is made in segments for convenience of transport, the pieces forming the complete ring being bolted or riveted together at the site of the foundations. The wooden curb is composed of several thick layers of hard wood planking cut to the proper shape, and laid with broken joints, the whole being bound together with suitable bolts and spikes. In some cases the lower or cutting edge of the wooden curb is strengthened or protected by a sheathing of wrought-iron plates.

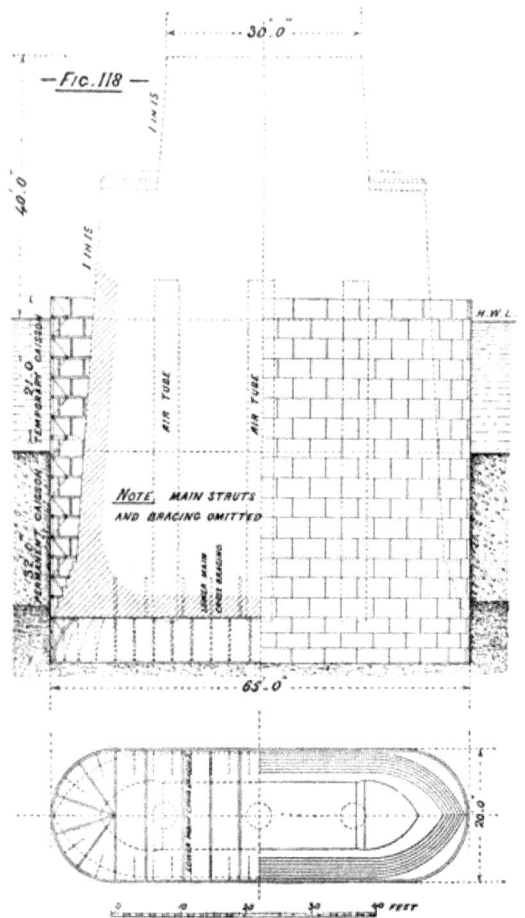

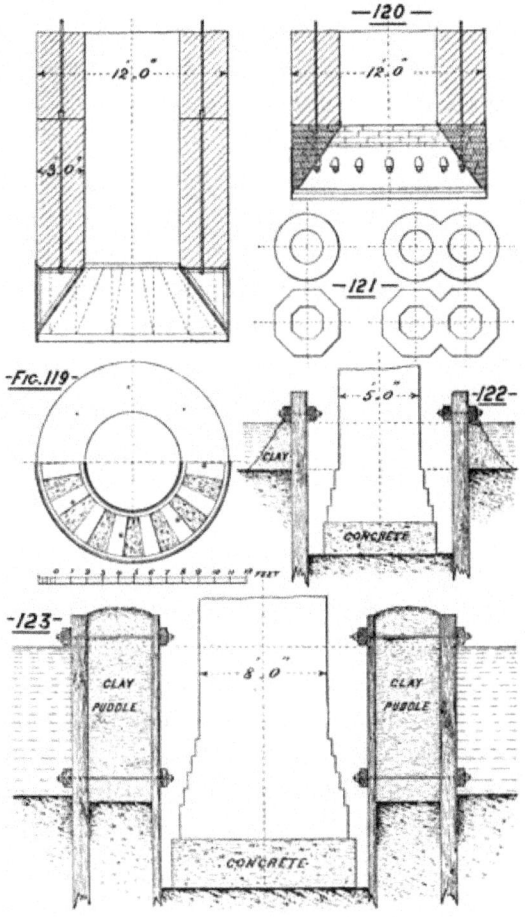

Well foundations are usually put in when the rivers are at their lowest, and reduced to a few small channels in the great width of dried-up river bed. This condition enables the greater portion of the curbs to be conveniently and accurately placed in position on dry ground, or on ground which, although soft and muddy, is not covered with water. Should the site of one of the wells occur in one of the small channels, the stream can be diverted to one side, and a small artificial island made to receive the curb above water-level. When a curb is fairly fixed in position, the work of building the brick well can be commenced. With the wrought-iron curb the triangular cavity between the vertical plate and sloping plate must be filled with concrete to form a level base for the first course of brickwork. The wooden curb being composed of horizontal layers of timber, is ready to receive the brickwork without further preparation. To strengthen and keep the brickwork firmly tied together, strong wrought-iron vertical tie-rods, 1¼ or 1½ inch in diameter, are generally built into the work—as shown in the sketches—at distances about four feet apart. The lower end of the bottom tier of tie-rods is secured to the curb, and the upper end passed through a strong wrought-iron plate-ring, which is continuous all round the brickwork. A long deep nut is screwed down over the top or screwed end of tie-rod until the plate-ring is down tight on the brickwork. The tightening nuts are made sufficiently deep to receive the lower ends of a second series of vertical tie-rods, which in like manner pass through another wrought-iron plate-ring on the next section of brick well, and the same arrangement is continued for the full height of the well. The lengths of the tie-rods will depend upon the lengths of the section of brickwork to be built at a time, and may vary from 10 to 15 feet.

As the work of building proceeds the curb and brick well will sink gradually into the ground, and down to a certain depth, varying according to the material of the river bed, the weight of the brick well itself will effect the penetration and lowering. Beyond this depth the lowering must be done by scooping or dredging the material from the inside of the well, and placing heavy weights of old railway iron or other convenient masses on the top. When one section or length of well has been sunk down, then another set of tie-rods are inserted into the deep nuts, and another section of brickwork commenced. The operation of lowering is rather tedious, as all the weights have to be hoisted up on to the top of the length in hand, and piled so as to leave space for lifting out the material dredged from the interior; and then, when the length has been lowered, all the weights must be removed before the brickwork can be resumed on another length. Where the river bed consists of soft material, the excavation inside the well can generally be effected by suitable dredges or scoops worked from the surface or top of brickwork. Should trees or other obstructive masses be met with embedded in the strata, it will be necessary to employ divers to remove them piecemeal out of the way of the curb.

When the brick well has been lowered down to the full depth, and is thoroughly bedded in a stratum of strong material, the test weights should be left on for some time to ascertain if there is any further sinking. After all the weights have been removed the bottom of the well can be dredged out clean, and the interior filled in with concrete to such height as may be considered necessary.

Brick wells must be watched carefully to ensure that they sink down in a perfectly vertical position. Any inclination away from the perpendicular must be corrected at once by means of guys and struts, the same as in sinking iron cylinders. The principal difficulty will be with the first 20 or 25 feet.

The diameter of the well will depend upon the weight it has to carry, and its height from river bed to under side of girders. The wells may be either circular or polygonal in section, and built singly or in pairs, as shown in sketches (Fig. 121).

Many piers and abutments of bridges in shallow or moderately deep water are built by means of coffer-dams of timber and clay puddle. The coffer-dam forms a water-tight wall round the site of the foundation, from which the water is pumped out, and the excavation carried down to the depth required. In very shallow water it is sometimes sufficient to drive only a single row of piles, and form a bank of good clay puddle on the outside, as shown in Fig. 122. In deep water it is necessary to drive a double row of piles, 3 or 4 or more feet apart, and fill in the space between with clay puddle, as shown in Fig. 123. The piles for coffer-dam work should be carefully selected, of good timber straight, and correctly sawn on the contact faces. Guide-piles are first driven in proper line and position round the intended foundation. To these strong horizontal double waling pieces are securely bolted, one on each side of the guide-pile, one pair near the top, and the other pair as low down as can be placed. The sheeting piles, which are lowered down between the horizontal waling or guiding pieces, are driven as close to one another as possible, being assisted in doing so by the sheet-pile shoe, shown on Fig. 124, which is made not with a point like an ordinary pile shoe (Fig. 125), but with a cutting edge slightly inclined, so that in driving the tendency of the pile is to drift towards the pile previously driven. Sometimes the outer row of piles consists of whole balks, and the inner row of half balks; the size of the piles must, however, be regulated by the depth and current of the water. When both rows of piles have been completed, the space between should be dredged out, and then filled with carefully prepared clay puddle. To enable the puddle to adapt itself thoroughly to the wooden sides, it is desirable to remove the inside walings after all the piles are driven, as any internal projections interfere with the proper punning and settling of the puddle. The swelling of the puddled clay has a tendency to force apart the two rows of piles, and to counteract this as much as possible, iron tie-rods should be passed through from side to side every few feet, and screwed up against large washers placed on the outside of the outer walings. Strong struts or cross-bracing of timber must be placed from side to side inside the coffer-dam to resist the pressure of the water in the river. This cross-bracing can be removed gradually as the work of building progresses upwards, and be replaced with short struts wedged in against the sides of the finished courses.

In cases where the ground is soft, and when it is not considered prudent to excavate the foundations deeper for fear of disturbing the stability of the coffer-dam piles, rows of large, square bearing-piles may be driven in the floor of the foundation, as shown in Fig. 111. The tops of these bearing-piles must all be sawn off to the same level, and a platform of strong double planking securely fixed to the piles to receive the foundation course of concrete, masonry, or brickwork. The spaces around the tops of the piles and the under side of the timber platform should be filled in with good cement concrete.

The interior of the coffer-dam is kept dry by constant pumping, either by hand pumps or steam pumps, according to the volume of water finding its way into the foundations. When the finished pier or abutment has been carried up above the river water-level, the coffer-dam is no longer required, and may be removed. Sometimes, to save the timber, the piles are drawn by means of strong tackle fitted up for the purpose; but in doing this there is considerable risk of disturbance to the foundations, and it is better to leave the piles in the ground and employ divers to cut off the tops a little above the bed of the river.

In preparing the design for a large foundation it is absolutely necessary to first ascertain by careful borings the description of material upon which that foundation must be placed, so as to proportion the area of bearing surface to the weight to be sustained. Some materials will naturally carry more weight than others, and although the engineer cannot always select the material he would prefer, he can, however, control the superficial area of the foundations. Much valuable information has been obtained both from experiments and from comparisons of actual practice, and the following memoranda may be useful for reference, as indicating the pressures per superficial foot which may be safely put on various materials:—

Moderately stiff clay	2½ tons.
Chalk	4 ”
Solid blue clay	5 ”
Compact gravel and close sand	6 ”
Solid rock	12 ”

Doubtless the above weights have been exceeded in many cases, but it is better to be on the safe side, and leave a good margin for stability.

Large subaqueous foundations for heavy piers and abutments are costly and tedious, and especially so when the pneumatic process has to be adopted. Special appliances and well-trained, experienced workmen are requisite, and if all the men and materials have to pass through the air-locks, the progress of the work must necessarily be slow. When the foundations have been completed up to the level of the water, the construction can be pushed on more rapidly, as the work of scaffolding, hoisting, and building, can all be carried on in the open air.

Amongst the very many types of arch-work and girder-work adopted for railway purposes, the following examples from actual practice may be useful for reference:—

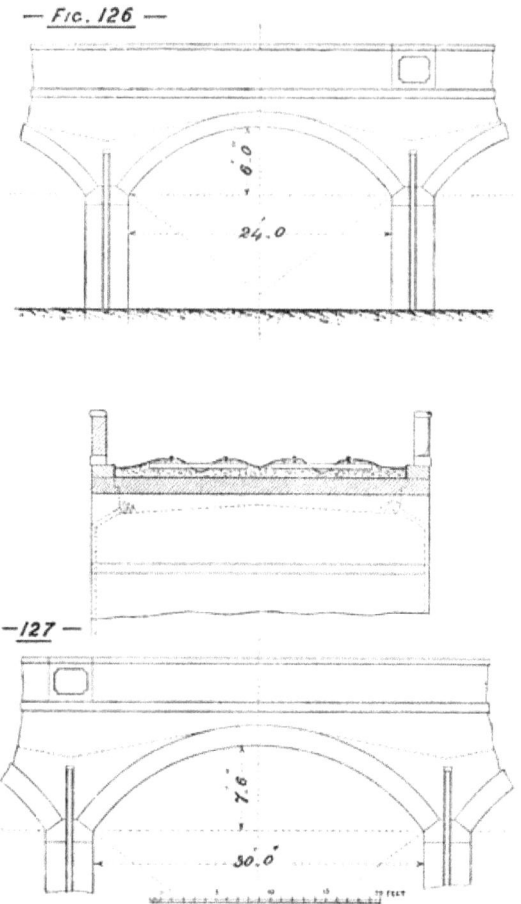

Fig. 126 represents small 24-foot span, low viaduct arching suitable for a line passing through towns or villages, where ground is valuable and the area to be covered must be kept as small as possible. The arches may be utilized for stables, stores, or roads of communication between the lands and properties intersected by the railway. The segmental form gives a better headway underneath than the semicircular, besides containing less material in the arching proper, and requiring a smaller amount of centering. Every precaution should be taken to prevent water percolating through any portion of the arching, or haunching, and a thick layer of good asphalte should be placed over the entire upper surface, and carried well up the lower portion of the parapet walls, as shown on the sketch. The cast-iron pipes with rose heads form a very efficient means of taking away the rain-water which filters through the ballast and filling. The pipes should be carried down in chases, or recesses, built in the fronts of the piers, to protect them as much as possible from injury in the yards below. Rose heads, pierced with holes, and surrounded with small stones hand-laid, serve well to conduct the water into the pipes. Where the arching is of considerable length, recesses or refuges for the platelayers may be obtained by substituting a short length of cast-iron-plate parapet, instead of the stone or brick parapet, over some of the piers, as indicated in the sketch.

Fig. 127 shows a similar description of arching for spans of 30 feet. The above two examples represent plain substantial work, but if circumstances warrant more external finish, this can readily be added without interfering with the general arrangement. In a similar manner, if considered preferable, the arches may be made semicircular or elliptical.

In the sketches shown of the arched over-line and under-line bridges, the arching and coping of parapets are in brick, and the remainder of the work in stone. In very many cases brick will be found cheaper and more expeditious for arching than stone, unless the quarries turn out stone in blocks which can be conveniently trimmed for arching. All bricks used for arch-work should be hard and well burnt, and special care should be taken in the selection of those to form the under-side course, which will be exposed to the atmosphere. For moderate spans arches have been successfully constructed of concrete. For this description of work the materials should be carefully gauged and mixed together, and the finished work should be allowed to stand some time on the centres to allow the concrete to become thoroughly set.

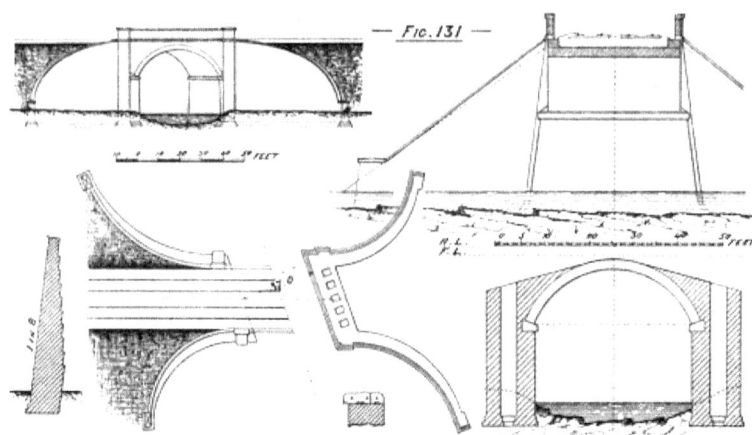
— FIG. 131 —

In Fig. 102, the cutting being deep, almost up to the level of the public road, the foundations of the wing walls are built in steps, resulting in a minimum of masonry below the finished ground line. Where the cutting is shallow, and the public road has to be brought up to the bridge on an embanked approach, the greater portion of the wing walls will have to be built up from the solid or original ground, and there will be a large amount of masonry below the finished ground line, as indicated in Fig. 128.

In some cases of over-line bridges it is necessary to curve the wing walls to correspond to the road which turns off to the right or left after crossing the railway, as shown in Fig. 129; or the wing walls may have to form two separate curves where the road branches off in two directions after leaving the bridge, as shown in Fig. 130.

Fig. 131 shows plan, elevation, and cross-section of an under-line arch bridge, considerably on the skew, carrying a railway over a river. The wing walls are curved, and very similar in type to some of those in preceding examples. The river bed and ground alongside being of solid rock, good foundations were obtained at a very moderate cost.

On many railways constructed in the beginning as single lines only, the over-line bridges have been built for double line. The additional cost in the outset has been small, compared with the great expenditure which would be incurred afterwards in reconstructing the bridges to suit a double line.

The general arrangement of abutments and wing walls shown in the foregoing examples will apply to similar classes of bridges where girder-work is adopted instead of arching.

There are many ways of forming the floor or deck of a girder bridge intended to carry a railway over a road or stream. In some cases it will be imperative to have a thoroughly water-tight floor to prevent rain-water percolating through to the roadway below; while in others, such as bridges over streams, and secondary roads, this special provision will not be necessary, and a lighter and more economical floorway can be adopted. A strong wrought-iron or steel-plate flooring, with its corresponding filling and ballasting, means not only so much additional cost in the flooring proper, but also so much additional dead weight to be carried by the main girders.

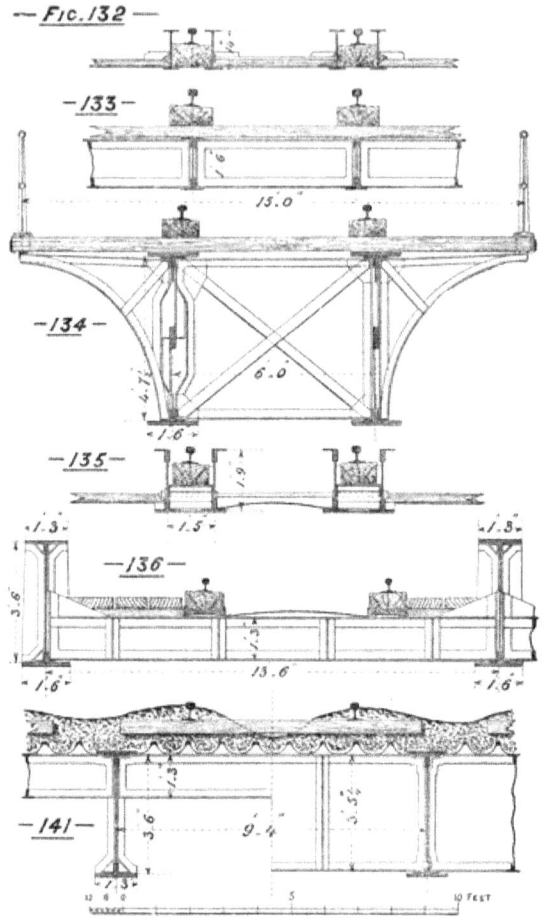

Fig. 132 is a sketch of rolled joist-iron **I**-girders and timber floor frequently adopted for small farm roads and cattle creeps of 10 or 12 feet span. A beam of timber is fitted in between the two rolled joist-irons, and the three pieces securely fastened together with strong iron bolts placed about 3 feet apart. These small compound girders rest on bearing-plates of wrought or cast iron, and are held together and to gauge by tie-rods, as shown. The rails are spiked or bolted down on to the timber beams, and the flooring formed of strong planking.

Fig. 133 shows an arrangement of plate girders for a 16-foot opening over a stream. The girders are placed immediately under the rails, and are tied together by plate-iron cross-bracing the same depth as the main girders. The flooring consists of 4-inch planking laid with ¾-

inch spaces, on which are laid longitudinal rail-bearers 14 inches wide by 7 inches thick.

Fig. 134 is a sketch of a somewhat similar arrangement for a lattice-girder bridge, 45 feet span, carrying a single line of railway over a river. The main girders are tied together by lattice-work cross-bracing. The floorway consists of 5-inch planking, laid with ¾-inch spaces, on which is placed the 14 feet by 7 feet longitudinal rail-bearers. Plate-iron outside brackets are riveted to the main girders to carry the ends of the planking and light tube-iron parapet.

Fig. 135 illustrates an example of trough girders, constructed to carry a double-line railway over a country road 25 feet wide, where the space from under side of girder to rail-level is small. The girders are constructed in pairs, with short, shallow cross-girders at 3 feet 6 inch centres, riveted in between them to carry longitudinal timbers on which the rails are laid. Bottom plates, 5/8 inch thick, unite the two girders for the length of their bearing on the abutments, and a similar plate, 9 inches wide, unites them at the centre; the remainder of the span is left open to prevent the lodgment of rain-water. Three strong tie-rods are placed to keep the girders to gauge. Curved wrought-iron ballast-plates are used between the running-rails, and plank flooring forms the rest of the covering.

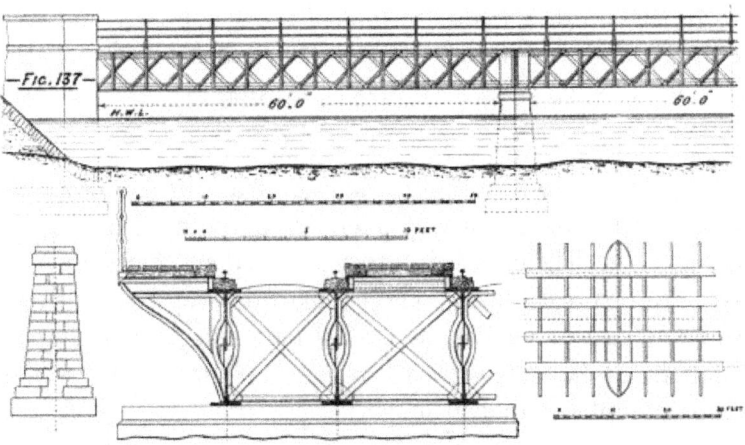

Fig. 136 is a sketch of a plate-girder bridge over a country road 28 feet wide, with the load carried on the lower flange of girder. Three main girders carry the double line of railway, the centre one having double the strength of each of the outside girders. On the top of the cross-girders, strong angle irons are riveted to serve as guides and supports for the longitudinal timbers which carry the rails. Every third cross-girder has raised ends to give increased lateral stability to the main girders. A close cast-iron plate parapet forms a screen to the roadway. Wrought-iron ballast-plates are used between the running-rails, and the remainder of the flooring is of timber.

Fig. 137 gives the particulars of one 60-foot span of a viaduct carrying a double line of railway over tidal water. The main girders are placed one under each line of rails, and all the four are strongly tied together by lattice-work bracing the full depth of the girders. The outside footpaths for the platelayers are carried on strong brackets, riveted to the main girders. Longitudinal timbers, coped with angle iron, are placed as outside guards, alongside each rail, for the full length of the viaduct. Wrought-iron ballast-plates are placed between the running-rails. The remainder of the footways consist of timber planking, laid with half-inch spaces, and covered with a layer of small pebbles as a protection against fire.

Fig. 138 shows a very similar arrangement in a viaduct carrying a single line of railway across a river. The two main lattice girders—66 feet span—are placed at 9-foot centres, to obtain greater stability. The cross-girders are extended to carry the outside footpaths and handrailing. Outside guards are placed alongside each rail as in the preceding example. Wrought-iron ballast-plates are fixed all along between the running-rails, and timber planking used for the rest of the floorway.

Fig. 139 gives cross-section of a lattice-girder bridge, 82 feet span, carrying a single line of railway over a river, with the load carried on the lower flange. The cross-girders are placed at 4 feet 3 inch centres. Wrought-iron ballast-plates compose the floorway between the rails, and timber planking covers the rest of the bridge. Plate diaphragms, or stiffeners, of the form shown at **A, A, A, A**, are riveted to the main girders at five places in their length.

Fig. 140 shows cross-section of a lattice-girder bridge of 200 feet span, carrying a single line of railway over a river, the load being placed on the lower flange. The floorway consists of plate-iron cross-girders, spaced at 4-foot centres, on which are placed the longitudinal rail-bearers and planking, the latter being covered with a layer of clean pebbles for the width between the running-rails. As the depth of the main girders was sufficient to admit of overhead bracing, strong plate-iron diaphragms, of the form shown on the sketch, were riveted to the main girders at every 50 feet. These diaphragms thoroughly brace the two girders together, and effectually prevent any tendency to side-canting, at the same time imparting an effective appearance to the bridge.

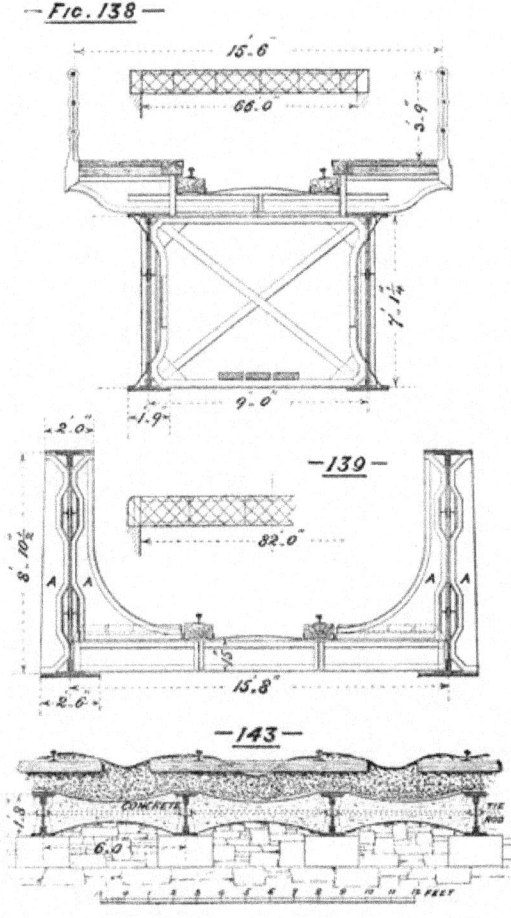

Fig. 141 shows cross-section of a plate-girder bridge, of 36 feet span, carrying six lines of way across a street. Strong plated cross-girder bracing, at 4 feet 8¼ inch centres, is riveted to the main girders, and the top is covered with old Barlow rails, 12 inches wide, and weighing 90 lbs. per lineal yard. A layer of asphalte, about 1½ inches thick, is carefully laid all over the upper surface of these rails to make a thoroughly water-tight floor. Clean gravel is placed on the top, on which are laid the sleepers and rails of the permanent way. Rain-water passes through the gravel into the hollows of the Barlow rails, and finds its way into suitable drains provided at each abutment. This arrangement not only prevents the falling of drip-water into the street below, but permits of the alterations of the lines of way, or putting in of cross-over roads on the surface above. The outside main girders are made deeper, and are surmounted by close cast-iron parapets.

Fig. 142 gives the particulars of a three-span plate-girder bridge, constructed to carry a double line of railway over two other railways and a canal, the load being placed on the lower flange. Two main girders are used for each line of way. Strong plated cross-girders are placed at 5 feet 3 inch centres, and on the top of these is laid a flooring of old Barlow rails, terminating at the sides with sloping wing-plates riveted to the cross-girders and main girders, the entire surface being covered with an inch and a half layer of asphalte. Good gravel ballast is placed on the top, on which are laid the sleepers and rails. One central main girder of sufficient strength would have been as efficient as the two central girders, but there was a practical difficulty which prevented its adoption. The new girder-work was built to replace an old structure of peculiar arrangement, and to keep the traffic going on one line there was no alternative but to make each line of way complete in itself.

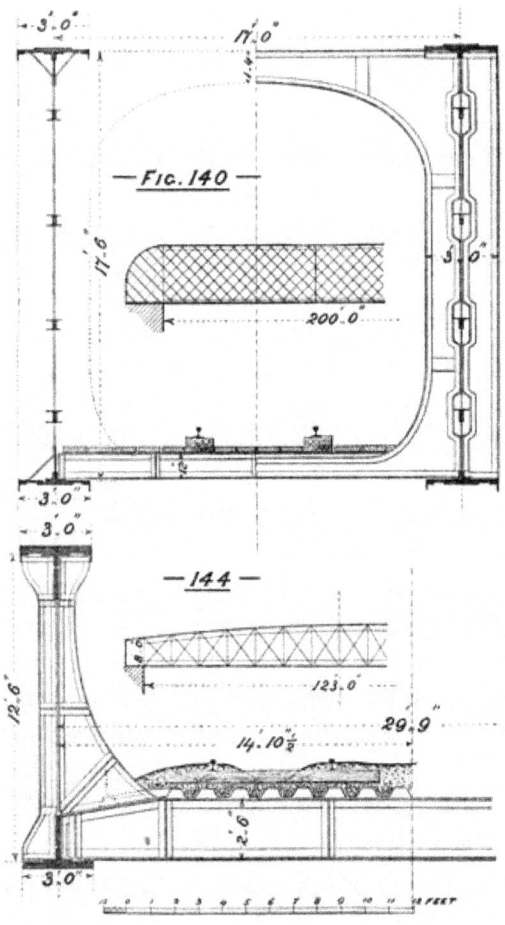

Fig. 143 illustrates an example of jack arches in concrete built between strong plate-girders. The span of the girders was only 16 feet, but the opening or roadway was of considerable length, and passed under a portion of a busy station yard. The girders are placed at 6-foot centres, and tied together in pairs by 1¼-inch tie-rods, three to the span, spaces of 6 inches in plan being allowed between each set of the rods. The concrete was curved up to the top plate of the girder, as shown, and the entire surface covered with a thick layer of asphalte, on which were placed the ballast and permanent way. Brickwork might have been used for the jack-arching, but concrete was considered more convenient.

Fig. 144 shows the cross-section of a truss-girder bridge of 123 feet span, carrying a double line of railway over a wide thoroughfare, the load being placed on the lower flange. There are two main girders, each 12 feet 6 inches deep in the centre, and 8 feet deep at the ends. Plate cross-girders are placed at 4 feet 6 inch centres, on which is riveted longitudinal plate-iron troughing, extending across the bridge and terminating at the sides with wing-plates, as shown. The entire floor is covered with a thick layer of asphalte previous to filling in with ballast to receive the permanent way. Plate stiffeners are adopted in this bridge very similar to those in Fig. 139.

Fig. 145 gives plan, elevation, and cross-section of a plate-girder bridge of 95 feet span, carrying a double line of railway over a very busy street. There are two curved-top main girders, each 10 feet 9 inches deep in the centre, and 6 feet 7½ inches deep at the ends. The arrangement of cross-girders, longitudinal plate-iron troughing, and permanent way, is very similar to that in the preceding example, but the side wing-plates are carried up higher, and are riveted up to the web-plate of main girder, forming continuous stiffeners from end to end of the main girders. A light, ornamental, close cast-iron parapet is bolted on to the top of the curved, or upper, boom of the main girder, the top line of the parapet being carried out parallel to the bottom boom of girder. This bridge crosses the street very obliquely, and, although cast-iron columns were allowed at the edge of the footpaths, the main spans are unavoidably large. When designing the above bridge, the writer had to adopt a girder that would form a screen, to provide a deck, or floor-way, which would be not only water-tight, but also deaden as much as possible the sound or vibration of passing trains, and at the same time give some ornamental appearance to the girders and parapets. This bridge carries a constant service of heavy trains; it is perfectly dry underneath, and is remarkably free from noise or vibration.

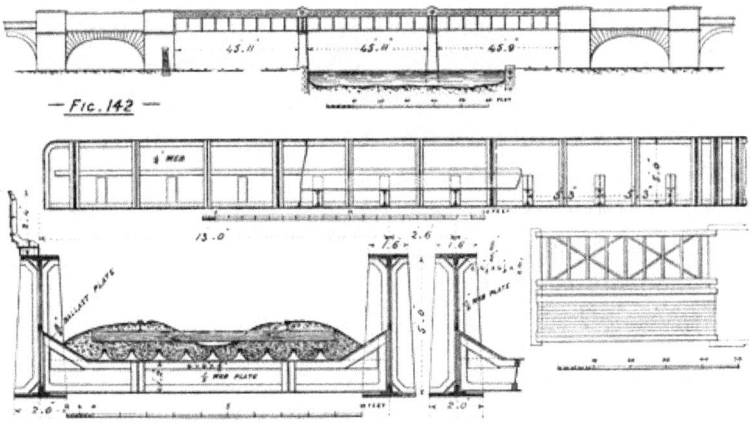

Fig. 146 shows cross-section of a plate-girder bridge of 40 feet span, carrying a double-line railway over a street, in a situation where the depth from top of rails to under side of girders had to be made as small as possible. Three main girders were used, the centre one being double the strength of each of the outside girders. Instead of ordinary cross-girders, transverse plate-iron troughing was adopted, very similar in section to the longitudinal iron troughing in Fig. 145, but stronger. The troughing rested on the angle iron of bottom flange of main girder, and was riveted to the vertical web-plates of main girders, shallow additional vertical plates being inserted alongside web-plates to prevent any drip-water or moisture coming in contact with the main web-plates. The entire surface of the troughing was well covered with asphalte before filling the hollows with gravel ballast. An ordinary transverse wooden sleeper was placed in each hollow, and on these sleepers the rails were secured as shown. In this case—as in others of transverse troughing—the rain-water had to be conveyed away from the hollow of each trough by a separate outlet into longitudinal gutters shown at **A**, **B**, and continued on to the abutments.

Transverse troughing is always more troublesome than longitudinal troughing, as both ends of each trough must be effectually closed to prevent the drainage water leaking out on to the web-plates, or angles of the main girders. With longitudinal troughing the water is readily carried away from each hollow, to cross drains constructed at the piers, or abutments.

Fig. 147 shows cross-section of a truss-girder bridge, 120 feet span, carrying a single line of railway over a river. The cross-girders are placed at 10-foot centres to correspond to the vertical members of the main truss-girder. Longitudinal plate-iron rail-girders are riveted in between the cross-girders, and the entire floor is covered with curved wrought-iron ballast plates, as shown. The rails are carried on longitudinal timbers, which are bolted on to the rail-girders. Angle iron brackets, riveted on the top of the cross-girders, keep the rail timbers in position and gauge.

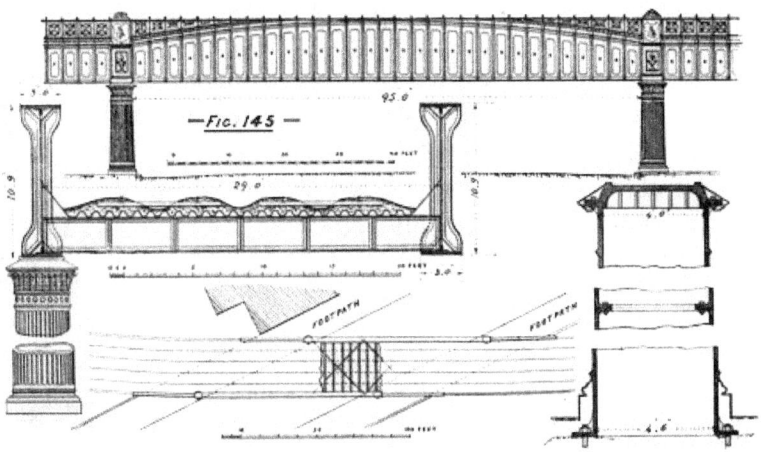

In each of the above examples, where longitudinal rail timbers are adopted, flange rails are shown, as many engineers prefer to have a continuous bearing for the rails on bridges, in case of rail fracture. There is nothing, however, to prevent the chair road being laid on longitudinal timbers, and for this purpose the writer has used chairs of the ordinary pattern, specially cast with side lugs to grip the timber, as shown in Fig. 148. Chairs of this form have a very firm hold on the longitudinal timber, and the side lugs check any tendency of the splitting or opening of the wood when putting in the spikes or screw bolts.

Fig. 149 shows cross-section of a plate-girder over-line bridge, 32 feet span, carrying a private road, 12 feet wide, over a double-line railway. The road traffic being small, the floorway was constructed of creosoted planking carried on rolled I-iron cross-girders placed at 3 feet 8 inch centres, and riveted to the main girders. The horse-tread track was provided with a second layer of planking, laid transversely, to take up the wear, cross battens, 4 inches by 2 inches, being placed at 12-inch centres, and sand spread between to give good foothold. A light lattice-work parapet was bolted on to the top of the main girders.

Fig. 150 gives cross-section of a plate-girder over-line bridge, 30 feet span, carrying a private road, 20 feet wide, over a double-line railway. The main girders are tied together by lattice-work bracing, spaced at 7-foot centres. Curved wrought-iron plates are laid across from girder to girder, and butt against a narrow horizontal plate, which forms part of the upper boom. The curved plates are riveted on to the top of girder, and form a continuous iron floor, or deck, from side to side of the bridge. Upon this iron floor is laid an ordinary asphalte roadway. The outside girders are made deeper, and carry an ornamental cast-iron parapet. In some bridges of a similar construction, the roadway is formed of creosoted wooden block paving, on a foundation of asphalte.

BRIDGE CARRYING THE D. W. AND W. RAILWAY (LOOP LINE) OVER AMIENS STREET, DUBLIN. [*To face p. 144.*

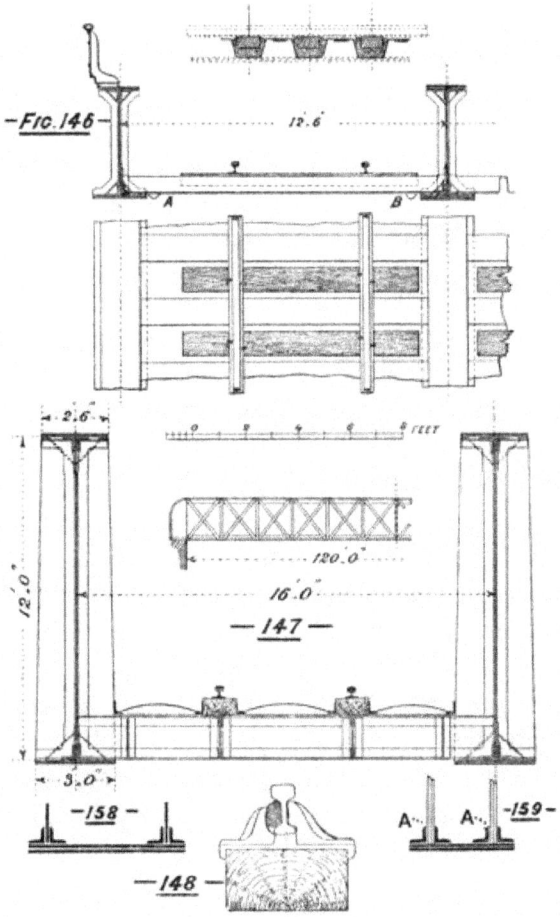

Fig. 151 shows cross-section of a plate-girder over-line bridge, 28 feet span, carrying a public road, 35 feet wide, over a double-line railway. The main girders, 2 feet 4 inches deep, are placed at 5 feet 2 inch centres, and are tied together by plate-iron cross-bracing 2 feet deep. Jack-arches of brickwork, 9 inches thick, are built in between the main girders, the haunching being filled in with concrete. The entire surface is covered over and made watertight with asphalte, on which is laid the metalling of the roadway. The outside girders are made considerably deeper, and have strong cast-iron-plate parapets bolted on to the top booms. There is no doubt that jack-arching of brickwork or concrete makes a very strong and permanent floorway, but its dead weight is very great, and its adoption is not to be recommended where iron or steel plate troughing can be obtained at a moderate price.

Fig. 152 gives cross-section of plate-girder over-line bridge, 41 feet 6 inch span, carrying a public road, 25 feet wide, over three lines of way. Two main girders are used, of sufficient depth to form parapets or screens for the finished roadway. Plate cross-girders, placed at 6 feet 6 inch centres, are riveted to the web-plate and lower angle irons of main girders; and on these is placed a flooring of plate-iron longitudinal troughing to carry the metalled roadway.

Fig. 153 gives the particulars of a plate-girder over-line bridge, carrying an important public road, 35 feet wide, over several main lines and sidings. The carriage-way is carried by two girders placed at 25-foot centres, and on the lower boom of these are riveted lattice-work cross-girders to receive the plate-iron longitudinal troughing and roadway. The footpath girders are set at a higher level, and the load placed on the lower flange. The curved side brackets merely act as bracing between the carriage-way girders and footpath girders. A cast-iron-plate parapet is bolted on to the top of each of the footpath girders, making a close screen, 6 feet high, above the footpath. Lattice-work cross-girders were adopted for the convenience of supporting small water mains and gas mains below the road-level. The roadway is formed of ordinary metalling, and the footpaths of asphalte pavement; the kerbing is of granite, and the side water-tables of crushed granite concrete.

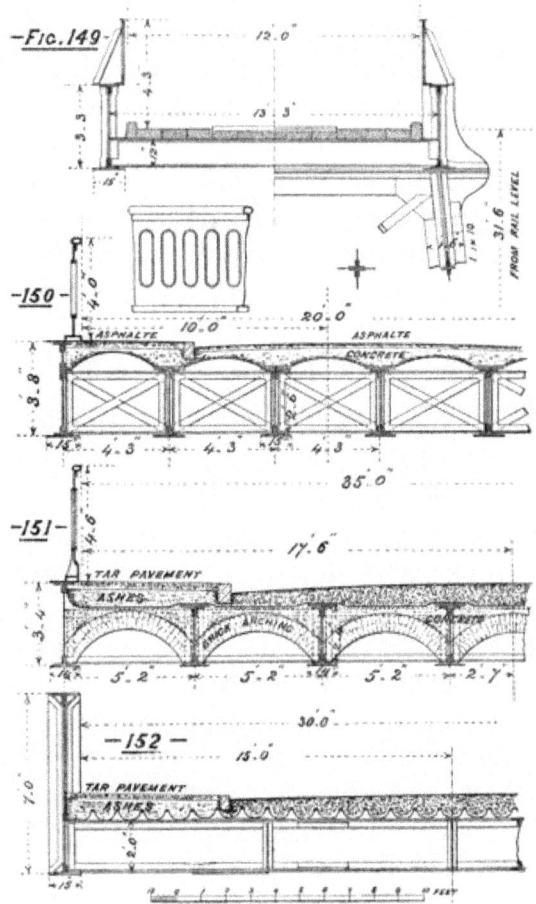

Fig. 154 is a cross-section of a small uncovered lattice-girder footbridge 41 feet span, and 5 feet wide, suitable for small roadside stations. The top and bottom flange consist each of two angle irons, those in the bottom flange being placed table side upwards, so as to bring the entire section of both angle irons fairly into play, and also to provide a better bearing for the channel-iron cross-girders which carry the planking of the footway. When planking is carried on the inside of light angle iron, as in Fig. 155, a severe strain is produced at the point **A**; this is entirely obviated by placing the bottom angle irons table side upwards, as in Fig. 156. Three of the channel-iron cross-girders are extended outwards, and to the ends of these are riveted tee-iron stiffeners to steady the main girders. In some cases stamped, or ribbed, wrought-iron plates are used for a footway, but, although more durable, they do not give such a secure or agreeable foothold as timber. The ascent or descent of the bridge may consist either of steps and landings, or of ramps, according to circumstances or expediency. Sometimes these bridges are made with curved tops, terminating in steps when nearing the steps, or ramps. It is very questionable whether such an arrangement is a good one or a safe one. There is always a feeling of insecurity when walking over a sloping surface broken up by steps, and experience points out that it is better to continue the footway level right across to the place where the passenger must change his direction to go down the stairs or ramp.

Fig. 157 gives cross-section of a covered lattice-girder footbridge, 62 feet 6 inches span, and 10 feet wide, suitable for an important station. The upper boom of girder consists of two angle irons and top plate, and the bottom boom of two channel irons. The cross-girders are rolled joist-irons resting on the top tables of the channel irons. Four of the cross-girders are extended outwards, and carry plate-iron outside vertical brackets to stiffen the main girders. Three-inch longitudinal planking is laid down from end to end of the bridge, and on this is laid 1¼-inch transverse flooring, in narrow widths, to form the walking deck. The footbridge is lighted from the sides by continuous glazed sashes fixed in strong wooden framework, as shown. The roof is covered with canvas bedded in white lead, and painted in the same way as an ordinary carriage roof.

The above examples of under-line and over-line bridges are given more with a view of illustrating some of the many different descriptions of flooring, rather than to point out or suggest the type of main girder to carry the load. The description and size of the main girders can be varied to suit the span of the bridge, the requirements of the traffic, and the opinion of the designer. For spans up to 50 feet it will generally be found that web-plate girders are both simpler and cheaper than lattice or truss girders; at the same time, there are occasions where plate girders can be advantageously adopted for very much larger spans, as, for instance, in the example given in Fig. 145, where the deep plate girders form a most efficient screen.

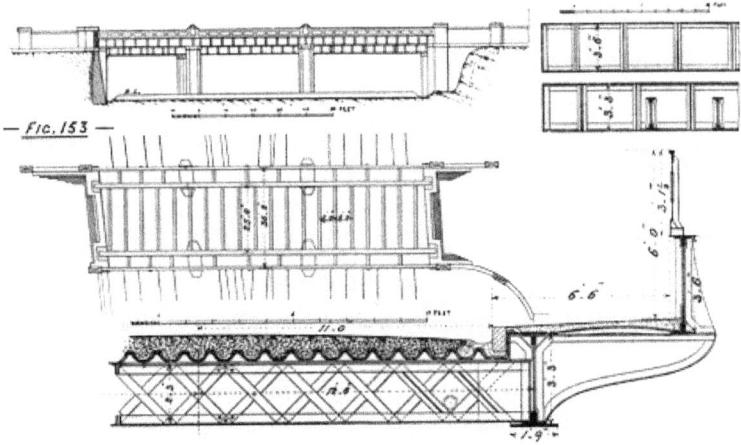
— Fig. 153 —

Figs. 160 to 194 give diagram sketches of a few out of the many forms of open, or truss, girders which have been adopted for large spans. There are many types from which to make a selection, each one possessing its own special features and advocates. In working out the details of any, or all of them, there are some points which should always be kept in mind when deciding the distribution of material in the main booms. Rain-water, or moisture of any kind, is the great enemy of wrought-iron or steel work, and therefore the plates, angles, tees, or channel sections, should be so arranged as to afford the least possible facility for the collection or lodgment of water. With open, level booms, as in Figs. 137, 139, 140, 144, and 145, the rain-water cannot collect, but runs off at the sides, and the plates are quickly dried by the sun and wind. With trough booms, as in Fig. 158, the collected rain-water can only get away through holes drilled for the purpose in the bottom plates. These holes are liable to become choked up, but even when open they rarely carry off all the accumulated water; some of it remains to corrode the plates, and is only dried up by evaporation. The inside of trough booms should be constantly inspected, and the exposed plates more frequently painted than the rest of the girder. In a similar manner, in small double-web lattice girders, with the lattice-bars inserted between two angle irons, as in Fig. 159, the rain-water finds its way into the spaces at **A, A**, in spite of the most careful packing or filling with cement or asphalte. Numbers of small girders of this latter type have had to be taken out after a comparative short life, in consequence of the great corrosion and wearing away of the lower ends of the lattice-bars and angle irons into which they were inserted.

It is most essential, also, that all portions of the girder-work should be conveniently accessible for inspection and painting. Complicated connections, and parts which are difficult to examine, are liable to be overlooked, or, at the best, only painted in a very imperfect manner. Neglected corners soon create deterioration, the paint scales off, corrosion commences, and the working section is gradually reduced. A discovered weakness in some of the important parts points to an early condemnation of the entire structure. The difficulty of access to the interior of box or tubular girders, especially those of small or moderate dimensions, is a great objection to that type of girder. Experience has pointed out that open girders, free and exposed to the light and air, can be so much more effectually inspected and painted.

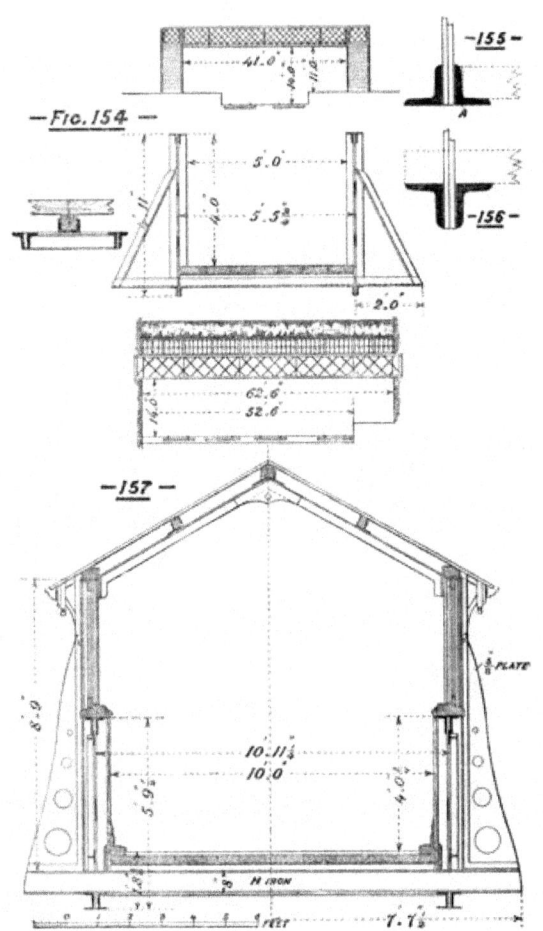

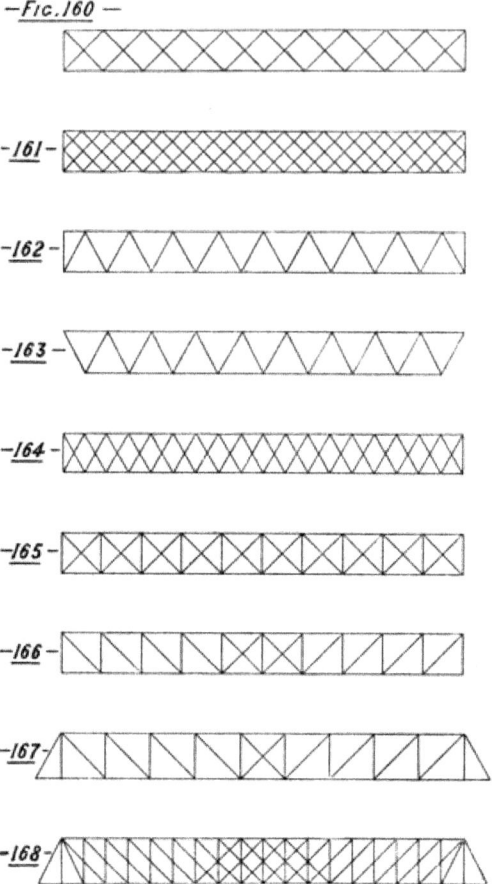

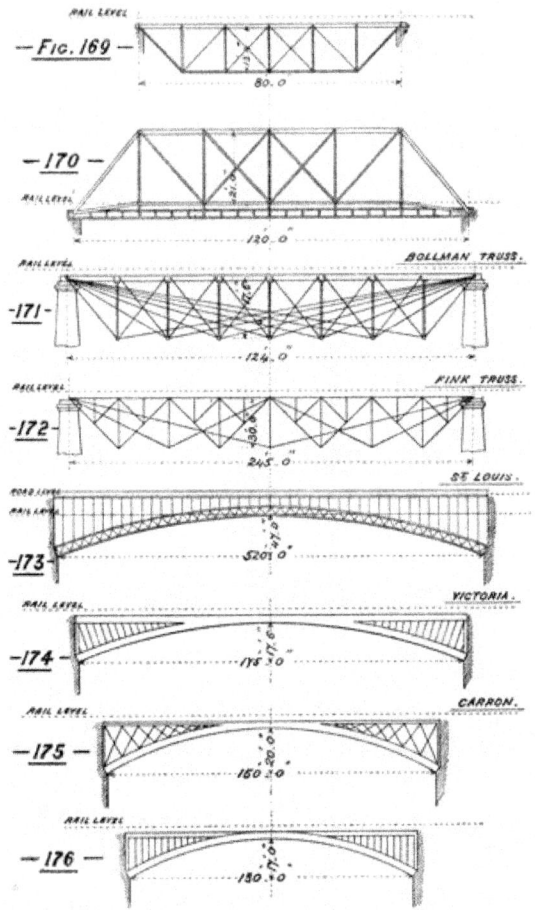

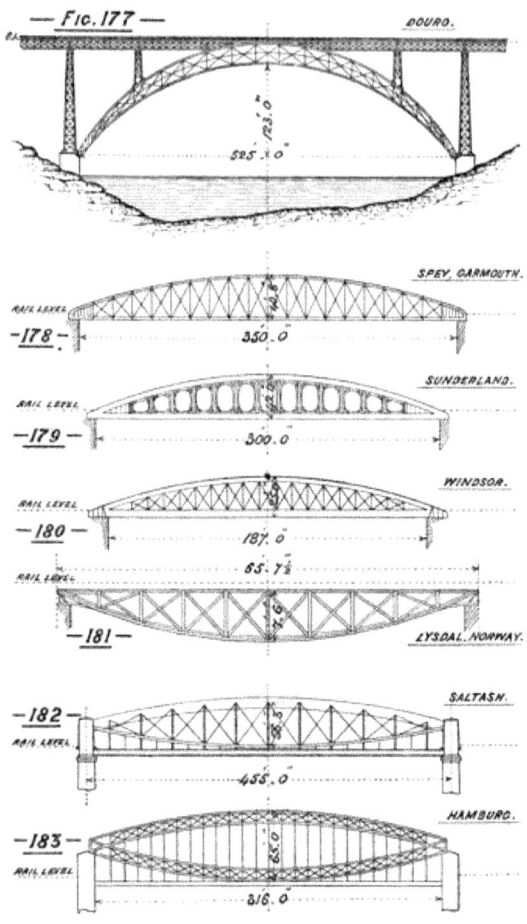

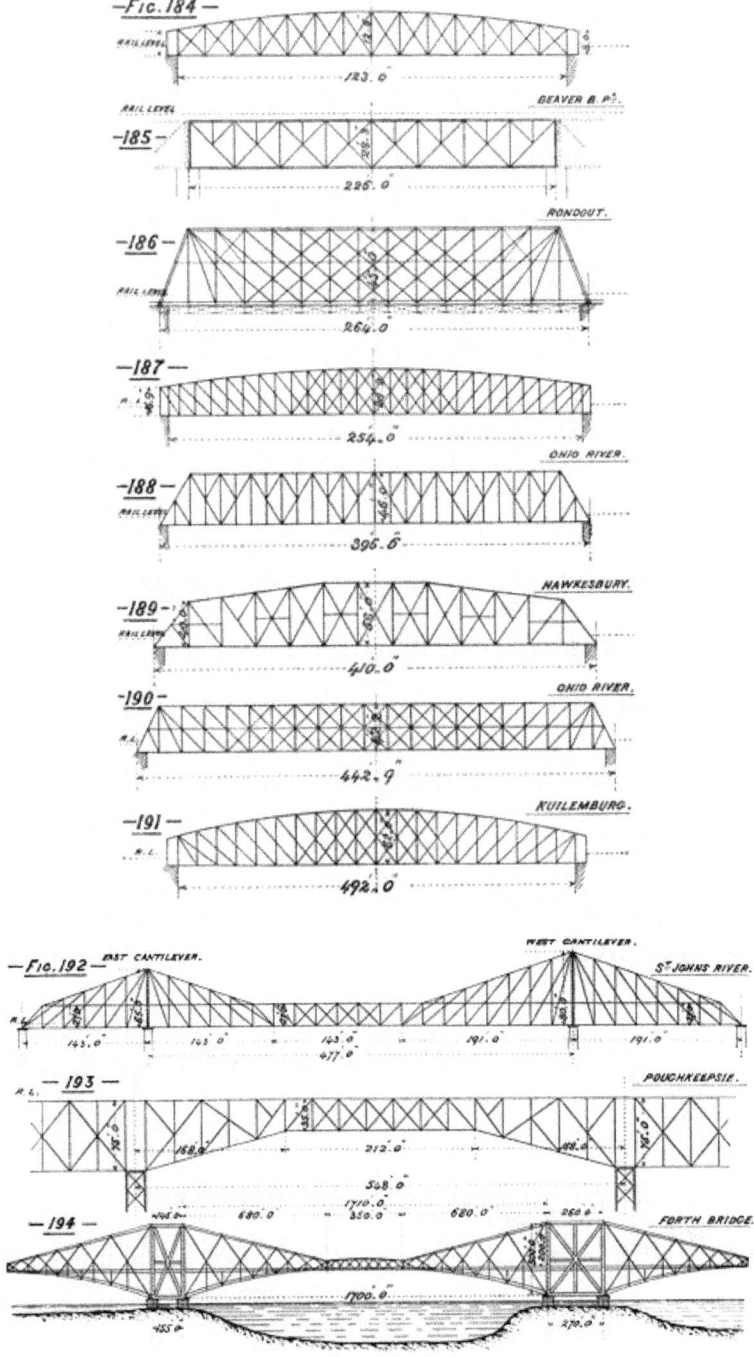

Perhaps one of the most anxious tasks which falls to the lot of an engineer is the renewal of under-line bridges and viaducts on a working line. On a new line in course of construction the entire site of the work is at the disposal of the erectors, and the building of a bridge or viaduct can be carried on with a freedom which cannot be obtained on an open line. On a working railway, the train service must be kept going, irrespective of renewals, and very often the best that can be done is to reduce the double line to single line working at the site of the operations. It is not always expedient or possible to make a temporary bridge and diverted line for traffic purposes, as the expenditure to be incurred might be too great to warrant the outlay, or there may be local difficulties to effectually prevent the introduction of a provisional structure. The taking down of one half of the old structure may necessitate the removal of stays and bracing affecting the stability of the half remaining to carry the traffic, and thus render temporary shoring and bracing necessary. The erection of the new work in such a limited space has to be watched with great care; all cranes, lifting appliances, and scaffolding must be kept clear of vehicles moving over the running-line, and very frequently it is found prudent to cease erecting operations during the passage of a train.

In very many cases of renewals, the description and arrangement of the old structure will materially influence or control the design for the new one, and the details of the latter must be schemed out so as to disturb as little as possible the stability of the old work remaining as the working road.

The following list gives the lengths of the main spans of some railway bridges, and may be found useful for reference:—

LENGTHS OF MAIN SPANS OF SOME LARGE RAILWAY BRIDGES.

Name.	Span.	Description.
	feet.	
Forth Bridge	1,710	Cantilever.
Niagara	821	Suspension.
Sukkur	820	Cantilever.
Poughkeepsie, U.S.A.	548	Cantilever.

Douro	525	Arch.
St. Louis	520	Arch.
Cincinnati	515	Linville truss.
Haarlem	510	Arch.
Kuilemburg	492	Lattice bow.
St. John's River	477	Cantilever.
Niagara	470	Cantilever.
Britannia	460	Tube.
Ohio River, Pennsylvania	442	Pratt through truss.
Saltash	434	Tube and girder.
Hawkesberry Viaduct	410	Compound truss.
Conway	400	Tube.
Vistula	397	Lattice.
Spey River, Garmouth, N.B.	350	Bowstring.
St. Laurence	330	Tube.
Hamburg	316	Double bow.
Cologne	313	Lattice.
Runcorn	305	Lattice.
Sunderland	300	Bowstring.
Rondout Bridge, Buffalo	264	Pratt through truss.
Newark Dyke (New)	259	Lattice bow.
Tay Bridge (New)	245	Lattice bow.
Ohio River, Louisville	245	Fink truss.
Beaver Bridge, Pennsylvania	230	Pratt deck truss.
Craigellachie Bridge	200	Lattice.
Rohrbach Bridge, St. Gothard River	197	Wrought-iron arch.
Windsor Bridge	187	Bowstring.

Victoria Bridge over Thames	175	Wrought-iron arch.
Shannon River Bridge	165	Bowstring.
Carron Bridge over Spey	150	Cast-iron arch.
Preston Viaduct	102	Cast-iron arch.
Trent River Bridge	100	Cast-iron arch.

Retaining Walls.—Instances frequently occur during the construction of a railway where it is advisable, if not absolutely necessary, to substitute retaining walls in preference to forming the slopes of cuttings and embankments.

The excavation of a cutting may be greatly reduced in quantity by introducing low retaining walls, as in Fig. 195, and the saving in the material to be removed will be all the more important in those cases where cutting is in excess of embankment.

The amount of filling for an embankment and the land on which it has to be formed may both be considerably diminished by building a low retaining wall, say 6 or 7 feet high, at the foot of the slope, as shown in Fig. 196. Such a retaining wall makes a most efficient fence and well defined boundary of property.

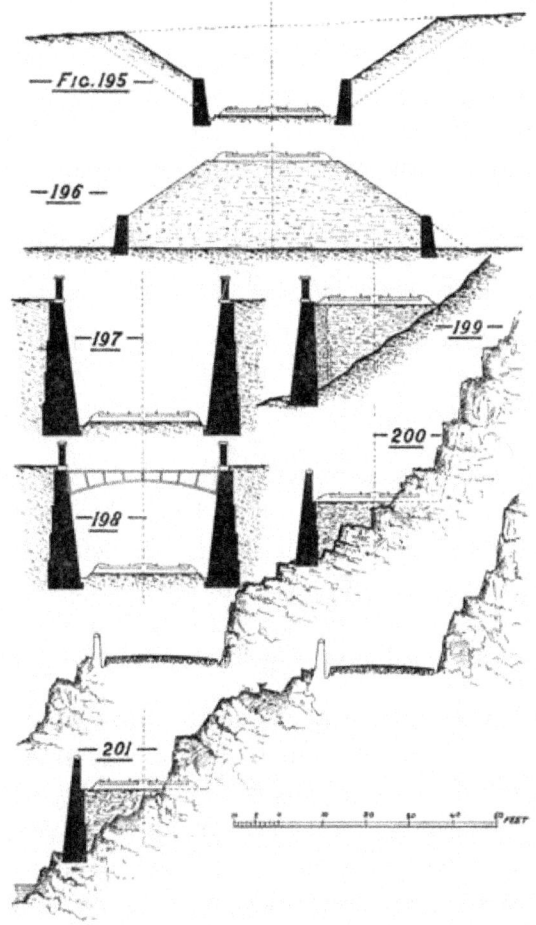

The policy of adopting low retaining walls in cases like the above will depend mainly upon the cost of building materials as compared with the cost of earthwork and land.

Where land is very valuable, and where residential property, streets, or roads must be interfered with as little as possible, the retaining walls may have to be carried up to the level of the original surface of the ground, as in Fig. 197, which is shown as for a cutting 25 feet deep. The walls may be built of masonry, brickwork, or concrete, or a combination of them, and the dimensions or thickness will depend upon the description of material to be supported. Weeping holes, or small pipe drains, should be formed in the walls, a little above formation level, to take away any water which may collect at the back.

Where the cutting is through soft, wet, treacherous clay, liable to slip or expand, it may be necessary to insert arched thrust girders extending from side to side, as in Fig. 198, so that the outward pressure against the one wall may counteract against the outward pressure of the other. The thrust girders should be placed at from 10 to 15 feet centres, and be well braced together in plan to enable them the better to resist any tendency of bulging out of the walls.

A similar arrangement of high retaining wall may be introduced in embankment to lessen the encroachment on streets or public roads, as shown in Fig. 199.

In making a railway through thickly populated towns, it is generally preferable to construct the line on arches rather than on earthwork filling between two high retaining walls. The numerous openings are available for future streets, or means of communication from one side to the other, and the arches themselves can be profitably utilized for stables, stores, offices, and workshops.

Fig. 200 shows a narrow rocky pass with deep rapid river on the one hand and high cliffs on the other, the only available ledge being already occupied by a public highway. By building a retaining wall, as indicated on the sketch, and excavating a little out of the cliff, space may be obtained for a line of railway; or the arrangement may have to be reversed, and the retaining wall for the railway built along the margin of the river, as in Fig. 201.

In both the cases, Figs. 200 and 201, not only must there be a number of weeping holes left in the lower part of the wall, but there must be sufficient well-built drains and culverts under the filling and through the wall to carry away all ordinary or flood water coming down from the cliffs and hills above. Where a retaining wall is built along the margin of a river, the lower portion, which will be in contact with the water when the river is full, should be constructed of selected large heavy stones to withstand the scouring action of the water, and any brushwood or floating timber which may be brought down by flood water.

Where retaining walls are built to support wet clay, or in embanked places on wet side-lying ground, the efficiency of the work will be much increased by constructing a layer, two or three feet in thickness, of dry, flat, bedded stones carefully hand-laid, from the foundation to the top of the wall, as shown in Fig. 199.

These dry stones form a continuous vertical drain to take away water from any part of the earthwork down to the outlets left in the lower portion of the wall.

The building of retaining walls entirely of dry stone is very questionable economy, and entails a constant expenditure in maintenance and renewal. The working out of one stone loosens the surrounding portion of the wall, and if not at once repaired, a length of the wall will fall down, bringing with it a large quantity of the earthwork.

If readily obtained, large heavy stones should be selected for the coping of retaining walls, so as to minimize as much as possible the chance of their disturbance or displacement. Where lighter stones have to be used, or bricks laid on edge, they should be bedded and pointed in cement.

In many places it is necessary to form wide and massive foundations of concrete on which to build the retaining wall; and in some cases of soft, treacherous ground, timber piling may be necessary.

Tunnels.—It would be difficult to assign a date to the first examples of subterranean works constructed for utilitarian purposes. Nature had furnished so many grand specimens of caves, grottoes, and underground passages formed in the solid rock, that man soon grasped the principle, and essayed to carry out similar works on his own account. The early attempts would probably be limited to forming places of shelter, storage or security. Advantage would be taken of those rocks which from their locality, accessibility, and compactness of material, promised favourable results. The appliances being few and primitive, the work of construction would be laborious and slow. So long, however, as the workers restricted their operations to the solid rock, they had merely to contend against the hardness of the material, as the opening or passage-way, once made, required no further support or attention; but as the wave of progress swept onward, man was compelled to deviate from the lines originally followed by nature, and had to form his subterranean pathway through softer material, where the workings required substantial support. The search for minerals of various kinds led to the driving of long headings or galleries underground, and as these had frequently to penetrate through strata of a soft and yielding character, strong timber framework had to be introduced to afford stability to the works, and safety to the workers. For ordinary mining operations, strong rough timber supports may meet all requirements, and may last until the heading is worked out and abandoned; but for subterranean passages or tunnels which are intended to form permanent means of communication, the strongest and most durable materials must be used to protect the interior as far as possible from deterioration or decay. Heavy timbering might be sufficient for mere temporary purposes, but substantial masonry or brickwork side walls and arching became necessary for permanent work in those portions where the tunnel required artificial support.

The first tunnels of any importance were most probably those constructed for canal purposes. Many of them were of considerable magnitude, and in some instances were from two to three miles in length. They were substantially lined with masonry or brickwork at all places where the tunnel passed through soft material or loose rock, and from the solid nature of the work, and the many years they have been in existence, they thoroughly testify to the ability of the constructors.

The introduction of railways involved the making of a large number of tunnels, perhaps more so in the beginning, when it was thought that the use of the locomotive would be confined to very moderate gradients, and when engineers hesitated to adopt the steeper inclines and sharper curves which form the practice of modern times. Another element of consideration also consisted in the fact that the first railways were designed to connect the most populous and busiest districts, where the prospects of heavy traffic would appear to warrant a large outlay for works of construction. As the system spread and railways extended further away from the important centres, the probabilities of traffic would become less promising, and efforts would be made to keep down cost of construction, and avoid tunnel work as much as possible.

It is not easy to define where cutting should end and tunnelling begin. There is no practical difficulty in making a cutting 50, 60, or 70 feet deep, with slopes to suit the material excavated, and the estimated cost per yard forward may even compare favourably with the cost of average tunnel-work. But there are other questions which must be kept in view—the time required to form the cutting, the space to be obtained on which to deposit the enormous quantity of excavated material, and the probable difficulty in obtaining the large area of land necessary for the cutting.

Before deciding the actual position of a tunnel, both as to line and level, it is necessary to obtain the most reliable information possible regarding the strata through which it has to pass. In addition to the geological indications on the surface and in the locality, borings should be made, and trial holes or shafts sunk along the proposed centre line of the work, and from these an approximately accurate longitudinal section can be laid down on paper, showing the respective layers of material to be cut through, and the angle at which they lie. With these particulars before him, the engineer may, in some cases, consider it more prudent to change the position of the tunnel in preference to incurring specially difficult or tedious work in dealing with some recognized unfavourable material. Occasionally the route may be slightly varied and better material obtained, but very frequently there is little to be gained except by a wide deviation from the original line.

Solid rock, except for the slow progress, is perhaps the most favourable material for tunnelling, as the timbering, side walling, and arching can be almost, if not entirely, dispensed with.

Loose rock, although more readily removed, necessitates strong timbering to prevent large masses breaking away and falling into the tunnel.

Some clays are very compact and tenacious, and will stand well with moderate timbering, but even these should not be left long before following up with the side walls and arching.

Many clays give much trouble by expanding, or swelling out, when the excavation penetrates the layer, and although extra strong timbering may be used, and be placed closer together, the logs and planks are frequently bulged out and broken by the action of the clay. Specially strong supports are required for this description of clay, and extra thickness of material in the permanent work of side walls and arching.

Solid unbroken beds of chalk are not difficult to cut through: the material is easy to work, and the excavation will stand with ordinary timbering; but where the chalk is broken and intersected with deep pockets of gravel and sand, the operations are very much impeded. The loose material, once set free by cutting through the confining barrier of chalk, will quickly fall into and fill up the excavation if not held back by strong timbering. Side walls and arching are generally necessary for tunnels through chalk.

Soft wet clay, quicksands, or other strata having springs of water percolating through them, are serious obstacles in the way of expeditious tunnelling. No sooner is one cube yard of this soft material removed than another slides down, or is washed down, to take its place. When once the excavation taps the water-bearing strata, large volumes of water will find their way into the workings, and must be conveyed away to the mouth of the tunnel, or pumped up through the nearest shaft. The timbering of the sides and roof through this description of working is very tedious, and attended also with a considerable amount of risk. The absence of really solid ground on which to place or shore up the supports, taxes the skill of the excavators, and very often, when a short length has been made apparently secure, it will come down with a run, compelling all hands to beat a hasty retreat. The permanent lining through such treacherous material should follow the excavation very closely, and special care should be exercised in building the walls, arching and invert.

In the excavation through stratified rocks it is necessary to note carefully the lie of the strata, whether horizontal, vertical, or shelving, as with each one the excavators are exposed to risks, against which every precaution should be taken. A large horizontal slab of solid-looking rock will suddenly break and fall down without any warning. A heavy mass from a vertical layer, perhaps unkeyed, or loosened, by an adjacent blasting operation, drops down when least expected; and pieces from the high side of the shelving layers detach themselves and slide into the working in a most unaccountable manner.

No attempt should be made to carry a tunnel through material which has been disturbed or at all affected by any natural slip or cleavage, as although the strata may be hard and compact in themselves, they have really no solid or fixed foundation. The sliding away, once initiated, is certain to continue, and, accelerated by the tunnelling operations, will most likely, sooner or later, crush in the tunnel and sweep away every vestige of the work. Amongst the great mountain ranges these natural disturbances are by no means rare, and it will be wiser to keep away from their locality, even at the expense of a longer tunnel. Unfortunately, instances are on record of tunnels made, or in course of construction, through hillsides which had already commenced to slide away from the more solid rock, and the ultimate results were a further sliding away and total destruction of the work.

The lower slopes and outlying portions of high mountains are the most exposed to these natural slips, and they should be most carefully studied before commencing any tunnelling operations through them.

To facilitate drainage, it is essential that a railway tunnel should be laid down with a gradient or gradients falling in the direction of one or both ends of the tunnel. In nearly all tunnels a considerable amount of water finds its way in through the weeping-holes left for that purpose in the side walls, and must be carried away in suitable drains. If the quantity of water be small, ordinary water-tables, one on each side, may be sufficient; but for large volumes of water it will be necessary to build substantial side-drains, or an ample culvert below the level of the rails.

The gradients in a tunnel should be moderate, and not by any means excessive, or likely to tax the hauling powers of the locomotives. When an engine is working nearly to the utmost of its power on a steep tunnel incline, and the speed has become very slow, the exhaust vapours or gases from the funnel strike the arching with great force, and are deflected down on to the footplate in such dense volumes as to almost suffocate the driver and fireman. The writer will never forget two or three trying experiences in foreign tunnels, when he and the engine-staff were compelled to leave the footplate and climb forward to the front of the funnel, leaving the engine to work its way alone. Except for very short tunnels it is wiser to have easy inclines, and to restrict the steep gradients to the open line, where the very slow travelling, or even the coming to a stand from "slipping," may not produce unpleasant or alarming consequences.

In tunnels of any length it is usual, where possible, to construct shafts extending from the surface of the ground overhead down to the tunnel below. These shafts serve the double purpose of enabling the excavation to be carried on at an increased number of faces, and act as permanent ventilators after completion. In some cases the shafts are sunk exactly over the centre line of the tunnel, in others a few yards away from the centre line. The latter arrangement, if not quite so convenient for hoisting material while carrying on the excavations, has certainly the great after advantage that anything falling or maliciously thrown down the shaft cannot strike a passing train. The short side-gallery, or space between the tunnel and the shaft, provides a good refuge for workmen employed in repairs, and a convenient site for storing a few materials advisable to keep on hand.

Occasionally favourable opportunities present themselves for making horizontal shafts. For a portion of its length the tunnel may be located at no very great distance from the precipitous sides of some deep mountain ravine, or run near to the cliffs on the sea-coast, and advantage can be taken to drive a lateral heading or gallery through which the material from the tunnel excavation may be conveyed and thrown out into the gorge or seashore below.

In many cases the surface of the ground rises so abruptly from the faces of the tunnel and ascends to so great a height, that shafts of any kind are entirely out of the question, and the whole of the work must be carried on from the two ends. The rate of progress is consequently much slower, and the ventilation more difficult. In a shaftless tunnel of considerable length, and with a frequent train service, the question of providing suitable appliances for promoting artificial ventilation is of the utmost importance.

When the centre line of the tunnel has been accurately set out on the ground, and the levels of the different parts of the work decided, the construction of the shafts and the driving of the headings can be commenced. Working shafts intended to serve for permanent ventilation are generally made nine or ten feet or more in diameter, and are usually lined with substantial brickwork or masonry. When the well-like excavation has been carried down a few yards, or as far as it can be taken without the risk of the earth falling in upon the sinkers, a strong curb of hard wood or iron of the same diameter as the finished shaft is laid down perfectly level and to exact position, and on this curb the ring or lining of brickwork or masonry is built up to the level of the ground. The first length finished, the excavation downwards is resumed, and the interior lining continued, either by allowing the first length to slide down as the material below is gradually removed, and building further lining on the top, or by excavating and propping up the curbing with strong timbers below. When working to the latter method, stout wooden props of convenient length, stepped on to sole-pieces, are adjusted to the under side of the wooden curb above, the material is then removed for the thickness of the brickwork or masonry, and another curb accurately set to level and position; on this is built a length of lining up to the first curb.

This work of under-building or under-pinning must be carried out with great care and in segments; no props must be removed until the curb immediately above is well supported by the new lining, and the inside of the lining must be watched and tested to prevent any tilting. All spaces at the back of the work must be filled in and well packed with hard dry material. As the shaft is continued downwards the mode of working may have to be varied; different descriptions of material may be encountered, and blasting, shoring, and pumping may each in turn be necessary.

When down to the full depth, the lower length of the shaft will have to be securely supported by strong timbers, until it can be properly built into and incorporated with the arching of the tunnel or side gallery.

The completion of the shaft enables the workings to be commenced on each side, the excavated material can be hoisted to the surface, and building material lowered down. When the tunnel works are finally finished, the lining of the shaft should be carried up until it is 15 or 20 feet above the level of the surface of the ground, and a dome-shaped iron grating placed on the top as a protection against stones or other articles which malicious persons might attempt to throw down the shaft.

Some shafts are only intended for the temporary purpose of lifting the excavations from below, or lowering building materials down, and when the work is completed they are filled in again and closed. These service shafts are generally made square in section, and are merely lined with wood. Strong vertical timbers are placed at the four corners, to which horizontal double cross-pieces are bolted, thick planking being placed vertically at the back of these cross-pieces to support the sides of the excavation.

The *heading* of a tunnel is a narrow passage or gallery cut through from end to end of the works in the direction of the centre line. Where there are shafts, the cutting of the heading can be pushed on from several points, and be completed much more rapidly than when the working is restricted to the two ends. Headings are usually made just sufficiently large for the miners to work, say about 5 feet 6 inches high by about 3 feet wide, the object being rather to expedite the driving of the driftway than to remove large masses of material. They must be set out with great accuracy, and be constantly checked as the driving is in progress. When completed from end to end, the centre line can be checked throughout, and the course actually taken compared with the course intended. If there has been much variation in the narrow pioneer pathway, either in line or level, the amount of the divergence must be rectified when ranging the final centre line for the full-size excavation.

Tunnels cannot always be delayed until the heading is cut through for the entire length. In many cases the heading, the full-size excavation, and the permanent lining have all to be carried on at the same time, but as the work of the heading is smaller in extent, that portion of the operations can usually be kept well in advance of the others. The critical moment arrives when the headings from opposite directions meet, as any deviation or want of coincidence must be adjusted in the portion of the tunnel still remaining to be opened out to full size. Some tunnels of moderate length have been constructed without any heading at all, the excavation being taken out to the full dimensions from the commencement.

The heading of a tunnel assists not only in the correct alignment of the work, but furnishes at the same time an accurate knowledge of the strata passed through. It is also of service for ventilation, communication, and drainage.

In some cases the heading is driven at the bottom of the tunnel section, as in Fig. 211, and in others at the top, as in Figs. 202 and 204. Many of the earlier tunnels were constructed on the former system, while of late years the latter method has been very largely adopted. The bottom heading may perhaps in some instances be more efficacious for drainage, but it is very liable to be frequently choked up when taking out the excavation to the full size, and the lower surface is much cut up by the movement and conveyance of materials. Another disadvantage arises from the necessity of removing such a large amount of the cutting approaching the tunnel entrance before a beginning can be made to the bottom heading. The top heading has the advantage that it requires less removal of open cutting previous to its commencement, and, being high up in position, there is less chance of its being stopped up by falling material, the finished excavations being carried out on the sides and below the heading.

Where the headings are cut through solid rock, stiff shale, or compact chalk, little or no supports are necessary, but where they pass through clay or loose material, timbering will be required for sides, roof, and floor. Rough round poles, about 6 inches in diameter, are generally used for verticals, and are firmly secured to transverse sole-pieces, and on the top of these verticals strong transverse top-sills are fastened by means of rough tenons or checks. Strong boards are inserted at the back of this framework to keep the earth from falling into the working. The distance apart of the verticals will depend upon the description of material excavated; in very soft places they will have to be placed very close together, but where fairly sound and tenacious they may be placed at about 3-foot centres. The excavated material must be conveyed away to the entrance of the heading in small hand-trucks running on planks or light rails.

The widening out of the excavation to the full size will be a repetition on a large scale of the work carried out in the heading, with the difference that, the exposed surfaces being of so much greater extent, extra care and precautions must be taken with the framework and shoring of the timbering.

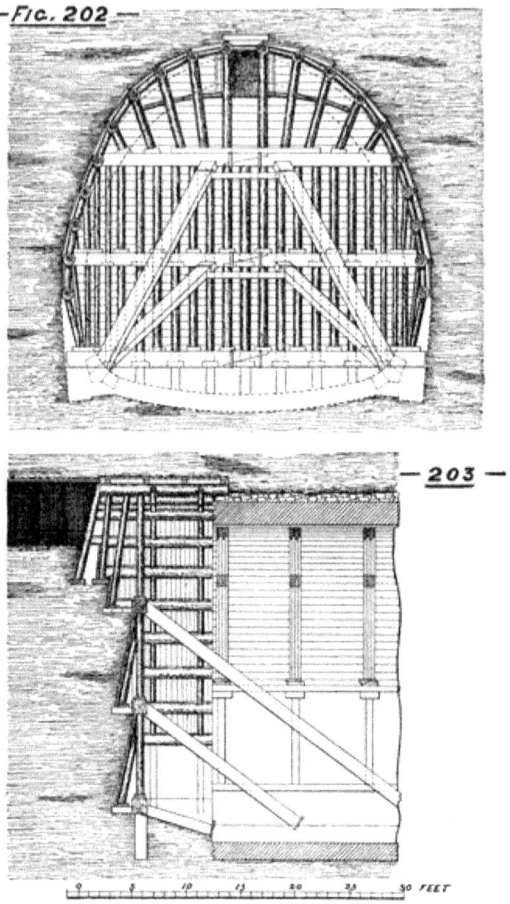

The form and arrangement of the timbering, as well as the number, sizes, and positions of the pieces, must be determined by the material of the excavation and the contour line of the finished arching or lining. The framework, which would be sufficient to support ordinary soft material, must be largely augmented both in quantity and scantling to meet the requirements for wet treacherous clay.

Figs. 202 and 203 give end view and longitudinal section of timber framework frequently adopted for average tunnel work. The positions of the different pieces will explain themselves and the duty they have to perform. The main struts, or raking pieces, which have to sustain great pressure, may be shored against the finished lengths of masonry or brickwork. The timbering of the sides can be removed as the lining proceeds, but in many cases the round logs and boards near the crown cannot be withdrawn, and have to be left in the work, the space between the top of the arching and under side of the boards being firmly packed with brickwork, masonry, or dry rubble stonework.

As the tunnel lining is generally carried forward in short lengths, following up the main excavations, the centering for the arching should be of such description that it can be readily transferred or moved forward as the work proceeds. The form of the centering, and the spacing of its upright supports, must admit of sufficient width for one or more lines of rails for the waggons required to remove the excavated *débris* and convey the building materials used in the lining.

Picks, bars, and shovels are the tools used in the excavation of the softer material and loose disintegrated rock, but for the hard rock, blasting will be necessary. The tunnel opening being comparatively small, only moderate blasting charges can be used with safety, and these must be placed so as to break up the rock-bed in a suitable manner for working, and without shaking or damaging the already completed excavation. Ordinary hand-drills, or *jumpers*, may be used for forming the charge holes, a number of them being at work at the same time, and the charges fired very closely one after the other. As the blasting operations necessitate the retiring of the miners to a considerable distance, out of the way of flying fragments, and the remaining away until the foul air has been dispelled, it is advisable to fire off several charges about the same time, and thus minimize as much as possible the stoppage to the drilling and clearing away the loosened material.

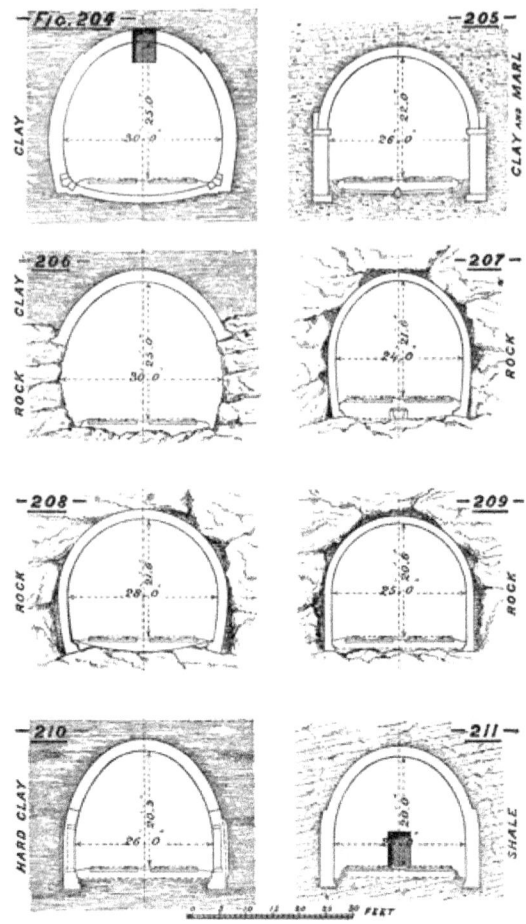

Mechanical drills, worked by compressed air or other motive-power, are now very extensively used where the rock is solid and continuous. They are much more expeditious than the hand drills, but they are costly in their installation, and also in their working and maintenance.

In some tunnels, where the material has been firm and dry, the upper portion of the excavation has been first removed, and the masonry and brickwork lining built in position down to about the springing of the arch, the remainder of the excavation being afterwards taken out, and the side walls built by means of shoring and underpinning.

In other tunnels the complete section has been excavated and timbered, and the work of building commenced from the foundation of the side walls. A strong continuous invert from side wall to side wall is necessary where passing through soft swelling clay or loose strata intersected with small streams of water. Where the material is very solid and dry, it is not necessary to introduce inverts, but the foundations of the side walls should be laid at such a depth below rail-level as not to be affected by drain-water running through the tunnel.

The side walls and arching may be either of masonry or brickwork, but should be of the best description, especially for the facework. For brick arching only the best hard-burnt bricks should be used, and the inner or exposed ring should consist of selected hard fire-bricks to withstand the heat and gases escaping from the funnels of the locomotives. The thickness of the side walls and arching will depend upon the description of material to be supported. In some places a comparative thin lining may be sufficient, while in others extra thickness must be given to resist the great pressure exerted by expanding clay and loose wet strata.

Weeping-holes, or small drain-pipes, placed low down must be left in the side walls every three or four yards, or closer in very wet places, to allow the water collected at the back of the walls to escape into the side drains of tunnel. In building the arch portion every effort should be made to have close solid work without any open joints or spaces through which the water may run, and the crown of the arch and a few feet down on each side should be coated with cement or asphalte to lead all water away from the top to the sides. Water dripping from the under side of the arch on to the line is a great destructor of the permanent way materials, especially the fastenings; and bolts, nuts, fish-plates, and spikes placed in a wet dripping tunnel will not last half the time they would out in the open line, where they would have the sun and wind to dry them.

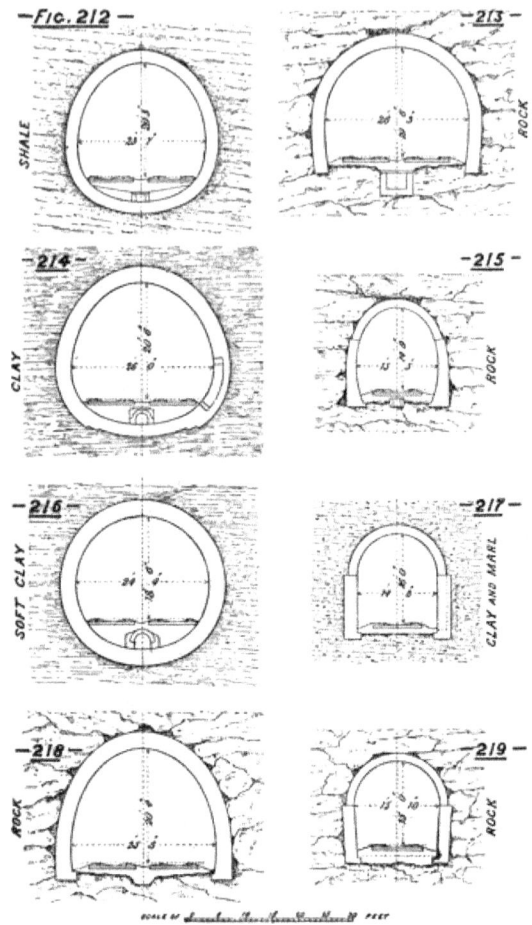

Small arched recesses or niches should be formed in the side walls at convenient distances to serve as refuges for platelayers or others working in the tunnels.

It is most essential that the space between the masonry and brickwork lining and the facework of the excavation should be carefully filled in and hard packed, so as to prevent the possibility of pieces of rock or other material falling on to the top of the arch. The neglect of this precaution may lead to a casualty years after the tunnel has been completed.

It would be impossible to over-rate the importance of a constant faithful supervision of the building of the lining, especially the arching. The work has to be carried on by workmen in cramped positions, with imperfect light, and surrounded by all kinds of obstacles and inconveniences, and unless a detailed inspection be rigidly maintained, a carelessness in the selection of the materials, and a laxity in the workmanship, will be the inevitable result.

Figs. 204 to 219 are sections of tunnels which have been constructed for double and single line railways. The sections give the normal form and dimensions adopted in each case, although there may have been many portions of the work where unfavourable or treacherous material necessitated an increase in the thickness of the side walls, or of the arching, or of both. The types vary in accordance with the opinions of the designers as to the most suitable section for the purpose, and range from the comparatively thin lining and vertical side walls shown on Fig. 207, to the almost circular form and very thick lining shown on Fig. 216. The latter is the section which experience has proved to be the best to sustain the enormous all-round pressure exerted by certain descriptions of swelling clay.

Careful judgment will be required to decide which parts of a rock tunnel may be left unlined. The apparently solid-looking portions are oftentimes deceptive, and numbers of instances are on record of large pieces of rock falling down in tunnels which for many years had been considered as thoroughly secure. Where there is any doubt it is better and safer to put in a lining, even if only to the extent of an arching springing from side walls of solid rock, as shown on Fig. 206. A moderate additional expenditure at the time of construction may prevent a serious catastrophe afterwards.

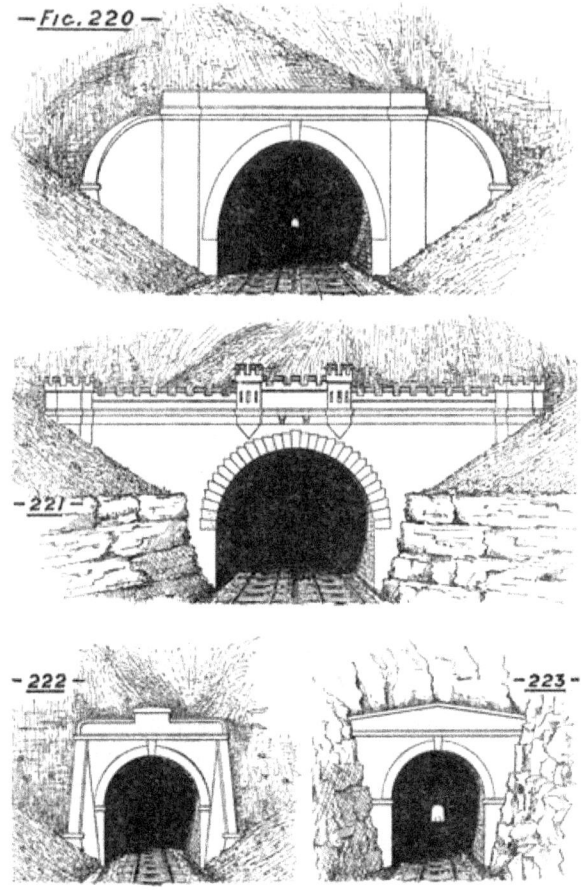

The faces or entrances to tunnels may be constructed with curved wing walls, as in Fig. 220, or with straight wing walls, as in Figs. 221, 222, and 223. Where the approach cutting is in rock, the latter form is generally adopted.

It would be misleading to put down any average price for tunnel-work. So much depends upon the locality, the description of material to be excavated, the cost of masonry or brickwork, and the cost of labour. Added to these come the unforeseen troubles of slips and water-laden strata, creating difficulties which baffle the miners for a time, and add enormously to the expenditure. Some tunnels for double line have been constructed in good ground, and under favourable circumstances as to building materials and labour, for as low as £32 per lineal yard; while others, carried out under adverse conditions, have cost as much as £150 per lineal yard. A medium somewhere between the two should represent the cost of tunnel-work through ground which does not present any special difficulty. At the same time it must be borne in mind that simple tunnelling which can be done in one locality for £50 or £60 per lineal yard, would be increased 20, 30, or 40 per cent. in another, where building material for the lining is scarce and expensive.

Tunnel-work abroad will generally cost more than the same work at home. The native labourers may perhaps be procured at low rates, but the skilled workmen must be brought from a distance, and will obtain high wages.

Another form of tunnel-work, generally termed the covered-way system, is frequently adopted in towns and places where land and space are very valuable. This method consists in the excavating and removing of earthwork to admit of the building of the side walls and arching of a suitable tunnel-way, and then filling in over the top to a depth of three or four feet, or up to the level of the original surface of the ground. This work may be carried out by either removing the entire width of the earthwork before the commencement of any building operations, or by first forming two deep, well-shored trenches, in which to build the side walls up to about arch-springing. In bad ground the latter arrangement has the advantage, as the shoring and strutting to hold up the sliding material is limited to the widths of the two narrow trenches, and the centre block of earthwork is left untouched as a support to the strutting. When the side walls have been built sufficiently high the upper portion of the centre block of earthwork can be removed to allow of the erection of centering and building of the arching, and afterwards the remaining portion of earthwork can be removed at convenience. In this manner a tunnel-way may be constructed under streets, gardens, and even under buildings. Being nearly all done in the open, the work is more under control than in an ordinary tunnel, but it is usually very costly. Temporary or diverted roads must be arranged; the excavated material must generally all be removed by carts, sometimes to long distances; and provision must be made for diverting the network of sewers, gas, and water pipes which are intercepted along the route.

Fig. 224 is a sketch of covered-way with brick arching. Fig. 225 illustrates another type where cast-iron girders and jack-arches of brickwork were introduced on account of the small headway. In soft yielding clay it is necessary to construct strong inverts, as indicated in the sketches. Recesses for the platelayers should be provided every ten or fifteen yards.

The above systems of covered-way were largely adopted in the construction of the underground portions of the Metropolitan Railway and District Railways in and around London.

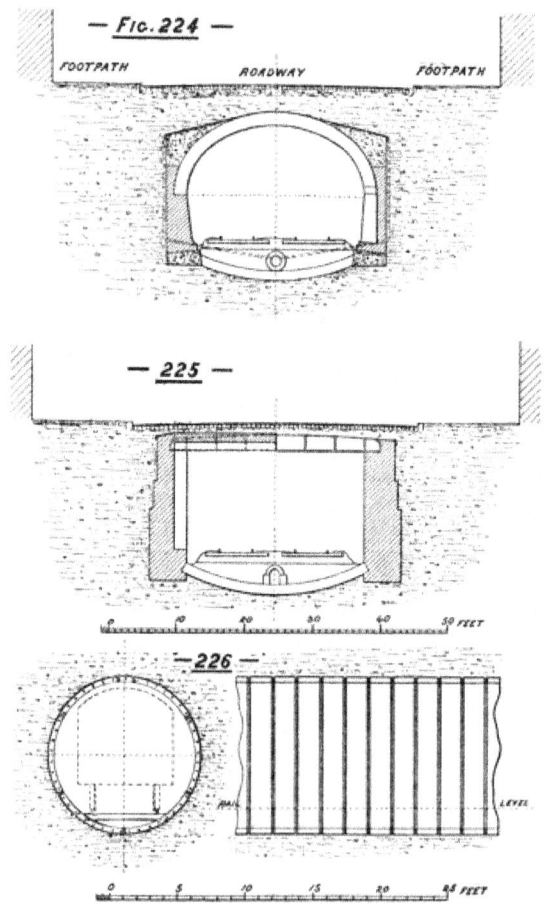

In addition to the ordinary type of tunnel formed by first excavating the material and then lining the opening with brickwork or masonry, tunnels of moderate size have been constructed of cast-iron tubes, similar in section to Fig. 226. The tubes were cast in short segments, bolted together inside, the outer circumference, or surface in contact with the earth or clay, being left free from projections of any kind. By making the segments with bolt-holes exact to template, they were readily fitted together in the work, and a thin layer of suitable packing material placed between the bolting-flanges sufficed to render the tubes water-tight. The tunnelling was carried on by means of a short length of slightly larger tube, or cap, made of plate-iron or steel, which fitted over the leading end of the main tube. The front end of this cap was made very strong, and provided with doors through which the miners could work. A series of hydraulic presses attached to the cap were brought to bear on the bolting-flange of the last completed ring, and as the excavated matter was removed by the miners from the front the cap was forced forward by the hydraulic presses, and another ring of cast-iron segments inserted. On the City and South London Railway, constructed on the above system, the small annular space formed round the cast-iron tube by the operation of the sliding cap was filled in with cement grouting by means of an ingenious machine designed for the purpose.

Large tunnels under rivers or tidal estuaries must each be dealt with according to the particular circumstances of depth below stream-bed, material to be cut through, length of tunnel, and gradient. The chief obstacle to be contended against in so much of the river tunnel-work is the large volume of water which pours into the workings through fissures in rock or seams of gravel and sand, necessitating the constant use of most powerful pumps. In ordinary land tunnels the gradients are generally laid out to fall towards one or both entrances, and any water finding its way into the excavations may be led away to the entrances by drains or pipes. On the other hand, in a river tunnel the gradients generally fall away from the entrances down towards the centre of the river, and all water coming in must be pumped out and raised up to at least the level of the river. In places where the water comes streaming in from many points, any failure or stoppage of the pumps would place the lives of the miners, and the security of the work itself, in great jeopardy. Iron shields, or protection chambers for the miners advancing the excavation, have been used with great success in carrying on work through loose wet strata which appeared to defy all other means of progress. Solid rock, chalk, or compact clay, may present no difficulty so far as they go, but a continued dip in the gradient, or a line of fault, may suddenly change the entire course of operations, and require the immediate use of the most powerful pumping machinery and protective appliances. The special features of each case will demand special precautions, and the judgment and inventive powers of the engineer will be severely tested in coping with the difficulties with which he is surrounded.

CHAPTER III.

Permanent way—Rails—Sleepers—Fastenings—and Permanent way laying.

Rails.—Accustomed as we now are to the substantial character of the permanent way of our railways, we can scarcely realize that in the earlier examples the rails or tram-plates were made of wood. The first lines of which we find any record were those constructed to facilitate the conveyance of coal, iron ore, stone, slate, or other heavy materials to shipping ports or points of distribution. Speed was a matter of little importance, the principal object being to introduce a distinct surface or roadway which would allow a heavier load to be hauled without increasing the hauling power. As a heavily loaded wheelbarrow, difficult to move on an ordinary road, can be readily wheeled along a wooden plank, so it may have been inferred that strong timber, laid in parallel lines and level and even on the upper surface, would form a track, or roadway, presenting far less resistance than the ordinary gravelled or paved roads.

The wooden tramway was the first improvement over the ordinary road. The idea once originated, various types were soon introduced, and the sketch shown in Fig. 227 illustrates one which appears to have been early suggested and largely adopted. Wooden cross-sleepers, **A, A**, were placed at convenient spaces, and on the top of these strong timber planks or beams, **B, B**, were spiked at proper distances to suit the wheels of the waggons or four-wheel trucks, which had flat tyres like ordinary carts. The spaces between the sleepers were filled in with gravel or broken stone to form a roadway or hauling path for the horses. A little later *double rails* were introduced, by placing a second or upper timber on the top of the lower one, as in Fig. 228.

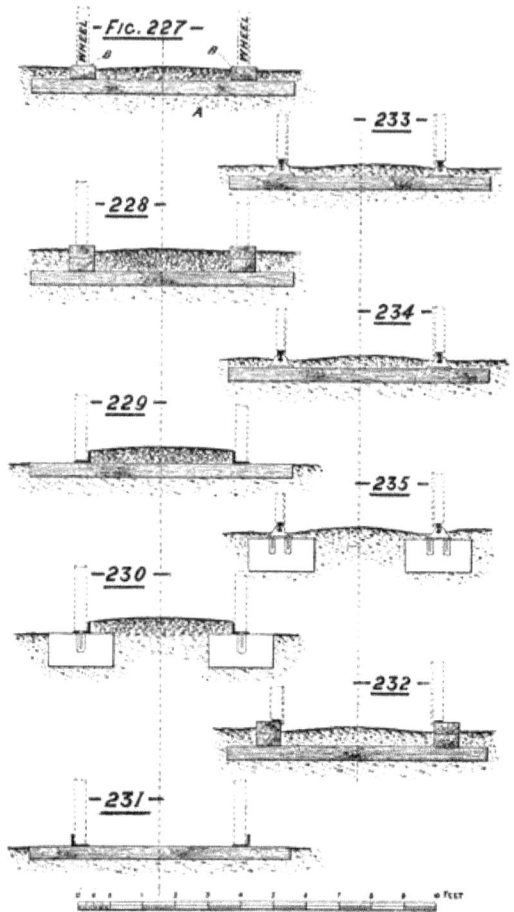

This double rail arrangement not only strengthened the framework, but by increasing the height allowed a greater quantity of suitable material to be placed over the sleepers to protect them from wear by the horses' feet. It can be easily understood that a wooden tramway could not be very durable. It would be affected by the sun, rain, and snow, and particles of sand and gravel thrown on to the tram beams from the hauling path would hasten the abrading or wearing away of the soft portions of the timber into hollows, leaving the hard knots standing out as projections. The uneven surface would produce a series of blows every time a loaded truck passed along, loosening the pieces and rendering the repairs constant and expensive. To obviate the rapid wear of the tram-timbers continuous narrow bars of wrought-iron were fastened on to the running-surfaces; these in a measure prolonged the life of the timbers, but at the same time added to the number of the pieces and fastenings to be maintained.

Primitive as this description of road appears to be, it was in use for many years in some parts of the United States of America, and even after the introduction of the early locomotives; timber was abundant and cheap, and iron in any form was costly. These long thin strips of iron, placed as in Fig. 232, had a tendency to become unfastened at the ends, and to curl up in a very alarming manner, which earned for them the soubriquet of *snake heads*. Although iron was only used to a limited extent in the first instance, it was soon found to be a much more suitable material for a tram-path than the best timber. As a next progressive step we find that the tram-plates were made entirely of iron, of full width for the wheel-tyres, and with a guiding flange to keep the wheels on the proper track. In some cases the guiding flanges were placed inside the wheels, as in Figs. 229 and 230, and in others outside, as in Fig. 231. With the former plan a thicker covering of gravel or broken stones could be laid down to protect the sleepers under the horse-path.

These solid tram-plates were made of cast-iron, that metal being considered the most convenient for manufacture and the least liable to suffer loss from rust and oxidization. Another advantage of the cast-iron was that broken tram-plates could be melted down and recast at a moderate cost.

Long lengths of these cast-iron plate tramways were laid down in this country and abroad, and short portions of some of them remain in existence even to the present day. They were of immense service for the transportation of heavy materials, and without their adventitious aid many valuable collieries and quarries must long have remained idle and undeveloped. In thus providing a level, smooth, and comparatively durable wheel-track for the waggons, these tramways became the fitting pioneers of the great railway system which was to follow.

Notwithstanding the great superiority of the cast-iron plates as compared with the former timber beams, much inconvenience was still caused by gravel and dirt falling on to the wheel-track and seriously impeding the haulage of the waggons. To overcome this difficulty the next step taken was to remove the guiding flange from the tram-plate and transfer it to the wheel, thus developing and introducing the original flanged wheel. This was a most important step, and paved the way for other improvements. The rails, or *edge rails*, as they were at first called, were made sufficiently high to allow ample space for the wheel-flanges to clear the ground, and were secured to cast-iron chairs placed on wooden cross-sleepers, or in some cases to stone blocks, as shown in Figs. 233, 234, and 235. The narrow top of rail, and its height above the horse-path, effectually prevented the lodgment of gravel or dirt, and the flanges on the wheels ensured a more even course. From the irregular and easily choked-up tram-plate, the system changed to the clean rail and properly defined track. Waggons could be hauled with greater freedom, and with less wear and tear to themselves and to the roadway.

At this time the use of the steam-engine was becoming more general, and a fine field was opened out for its application as a motive-power on the tramways. Stationary engines, or *winding engines*, as they were called, were first employed to haul the trucks by means of long ropes passed round revolving drums, and supported at intervals by grooved pulleys placed between the rails at suitable distances. In this way fair loads could be conveyed, and at moderate cost; but the system was found to be only suitable for short distances, and it had the great drawback that horses or other motive-power were still necessary for sorting or distributing the trucks before and after their transit by rope haulage.

The next great advance was to place the steam-engine on wheels, to enable it to haul and accompany the trucks. Crude and imperfect as the primitive locomotives must have been, a very short trial of them served to show that the rails of cast-iron then in use were totally unfitted to form a trackway for the newly invented machines. The short fish-bellied cast-iron rails were made in lengths merely to extend from chair to chair; they possessed little or no continuity, and from the inherent brittleness of the material they were constantly breaking and giving way under the increased weights imposed upon them. It became necessary to adopt a more reliable material, and attention was naturally turned to forged or wrought iron. The suggestion once made was promptly responded to by the iron makers. Special machinery was designed and constructed, and very soon wrought-iron rails were manufactured in large quantities. At first they were made very similar in section to the fish-belly cast-iron rails, but in lengths to extend over three or four sleepers. The increased length gave greater stability to the road, and permitted an increase of speed. The manifest superiority of the wrought-iron rails led to their universal adoption, and a great impetus was thus given to their manufacture. Improvements were made in the machinery for rolling, and more care was bestowed in the working of the iron. Changes were made in the section of the rails; the fish-belly form was discarded, and a double-head type was introduced to give more lateral stiffness. At this period in its history the capabilities of the *iron road* began to be more fully recognized, and the supporters of the system foresaw a great future success, both for the conveyance of passengers as well as goods. Hitherto the tramroads or railroads had been used for minerals and merchandize only, but it was now claimed that on a carefully constructed line, and with improved locomotives and rolling-stock, it would be possible to convey passengers more conveniently and rapidly than by any other method.

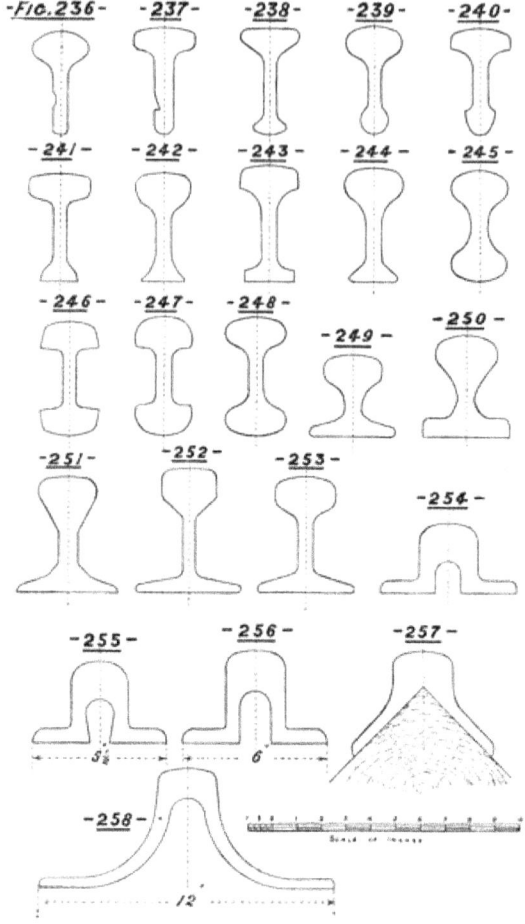

Inventive minds were at work to accomplish so desirable an object, and public enterprise was forthcoming to provide funds for the purpose. The successful working of the first passenger line formed the dawn of a new era in travelling, and similar lines were soon projected for other places. The wrought-iron rails in use at this time were generally of a double head form, and rarely exceeded 12 or 15 feet in length. They were held by wooden pegs in cast-iron chairs, which were secured to timber cross-sleepers or stone blocks, as shown in Figs. 234 and 235.

They were light in section, and it is stated that the first rails on the Liverpool and Manchester Railway weighed only 33 lbs. per yard.

The railway system spread rapidly, and the constantly increasing traffic of all kinds soon necessitated heavier rails. Various sections were devised and tried on different lines, one of the main objects in view being to obtain a steady road for the increasing speeds, as well as one of durability. Some of these sections are shown in Figs. 236 to 258.

Sections 236 to 248 all required chairs to attach them to the sleepers. The flange rails, 249 to 253, and bridge rails, 254 to 256, also rail 257, were designed to rest direct upon the sleepers without the necessity of chairs; and the Barlow rail, 258, with its great width of 11 or 12 inches, was intended to be used without sleepers of any kind, the gauge being secured by means of angle iron tie-bars.

Rails were rolled heavier and longer, and more care was bestowed on the fastenings; but, notwithstanding these improvements, the rail-joints still continued to be the weak point in the road. Even with an extra large joint-chair and stout wooden key, there was much vertical play at the ends of the rails, producing objectionable noise and vibration in the running, and acting detrimentally on all the fastenings. The introduction of fish-plates at the rail-joints, as shown in Fig. 259, effected an improvement which cannot be overrated, as by their adoption such security, speed, and smoothness became attainable as were not before possible. With a pair of simple rolled wrought-iron fish-plates, or splices, and four bolts—two through the end of each rail—a better, smoother, and more effectual joint was obtained than had ever been produced by the heavy cast-iron joint-chairs. The system of fishing, or splicing, was at once admitted to be the simplest and most direct method of joining the rails; and, although minor detailed improvements have since been made, the arrangement, as a principle, has never been superseded. Many miles of fished rails were laid down with a chair, or support, placed immediately under the joint, forming the method termed the supported fish-joint; but experience proved that this mode of application did not give such a good result as the suspended fish-joint, and the latter plan has now been adopted on almost all railways.

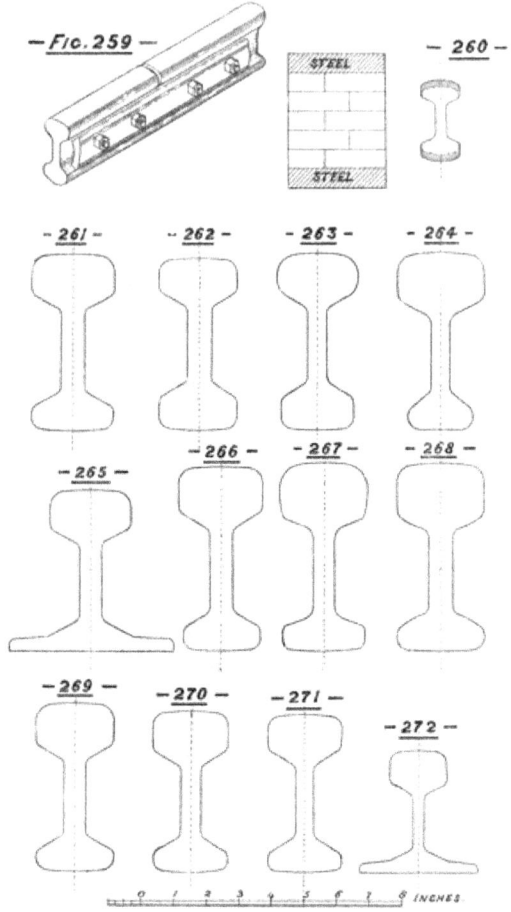

The experience obtained year after year in the wear of rails under heavy traffic, led to continued improvements both in the method of rolling and in the selection of the iron to form the rail-pile; one description of iron was found more suitable for the head, or running surface, and another for the vertical web; but, even with the best machinery and most carefully assorted materials, high-class wrought-iron rails were liable to lamination, and long thin strips of iron became detached from the upper, or wearing, surface. The rail was composed of many layers of iron, and it was not always possible to ensure that they were all thoroughly welded, or incorporated together. As early as 1854 a few experimental solid steel rails were laid down on some of the principal railways, and gave excellent results as to evenness of wear and durability, but their cost of manufacture rendered their extended use almost prohibitory.

Compound rails of steel and wrought-iron, as in Fig. 260, were also tried on several railways, but the practical results were not such as to lead to a very extended adoption. In preparing the *pile* for a compound rail, suitable wrought-iron bars were placed to form the lower member or flange, the web, and part of the head, and a slab of steel was placed on the top to form the upper portion of head, or wearing surface of the rail. It was intended that in the process of rolling these distinct layers were to be incorporated together, to form the section shown in Fig. 260. Doubtless many good wearing rails were manufactured on this system, but the inherent difference of the two materials, steel and iron, rendered it very difficult to ensure such uniform incorporation as would withstand the constant pounding under heavy, fast traffic. It was not until some years later that the process of the Bessemer Converter was discovered and perfected, by means of which steel can be produced in large quantities far more rapidly and at much less cost than by any other method hitherto adopted. The introduction of this process for making steel caused a complete revolution in the material for rails. Steel which had previously been excluded on account of its cost, could now be supplied at a moderate price, and, from its compact and homogeneous character, promised a very much longer wearing life than the best wrought-iron rails that had ever been rolled. Experience has shown that these promises have been fully verified; wrought-iron rails are things of the past, steel rails have taken their place, and can now be purchased at a less price per ton than the iron rails of twenty years ago.

It is interesting to note that out of the many varied sections that have been designed, some of which are shown in the sketches described, only two have practically survived—the bullhead rail and the flange rail. The bull-head rail, Fig. 261, has grown out of the original double-head rail, which had both the top and bottom members made to the same section and weight, with the object that, when the upper table had become so much worn as to be unfit for further use, then the rail could be turned, and the other table, or head, brought into service. Experience, however, proved that turned rails formed a most uneven and unsatisfactory road, the long contact with the cast-iron chairs resulted in serious indentations at the rail-seats, rendering the rails totally unfitted for smooth running. In practice, therefore, it has been found better to restrict the running wear to one head only, and to give increased sectional area to that head, and, at the same time to diminish the sectional area of the lower member to a corresponding extent, but to retain the same width, so as to obtain a full bearing surface on the cast-iron chair. Steel bull-head rails are now adopted on nearly all the principal lines at home, and on several of the leading lines abroad.

The flange rail, Fig. 265, was designed to give a broad, direct bearing on the sleepers, and thus avoid the necessity of using chairs. Rails of this section have been laid down on many of our lines at home, and are very largely used on the Continent, in the United States of America, and in our colonies generally. This section is, also, nearly always adopted for narrow-gauge railways. Having fewer parts, it makes a cheaper road than the bull-head rail, but is not considered so strong or suitable for heavy and fast traffic. Comparing the two rails shown in Figs. 261 and 265, each having exactly the same size and sectional area in the head, it will be seen that there is more material in the lower member, or flange, of the one rail than there is in the lower member of the other; the weight per lineal yard being 79 lbs. for the former and 75 lbs. for the latter. But this small excess in the weight and cost of the flange rail falls very short of the cost of the cast-iron chairs and wooden keys necessary for the bull-head rail.

Up to the years 1870-1875, it was the common practice to make the top, or wearing surface of the rail, comparatively round, as shown on the typical sections, Figs. 263 and 267. The effect of this sharp-curved outline was to limit the first wearing, or contact surface to a narrow strip along the head of rail, causing a tendency to groove or form hollows in the treads of the wheel-tyres. As the rail wore down, the upper surface assumed a much flatter curve, more closely assimilated to the section of the wheel-tyre, and giving better results for regular wear under heavy traffic. Profiting by this experience, the rails of the present day are made much flatter on the head than they were formerly, as will be noted from the sections shown on Figs. 261, 266, and 269, which represent types of rails now actually in use on some of the principal railways.

In designing a rail for any given line, the section and weight of the rail must necessarily be influenced by the weight of the rolling-stock passing over it, and the amount of the traffic it has to sustain.

The engine, being the heaviest vehicle in the train, will give the measure of the greatest weight on one pair of wheels. Engines vary considerably on different lines, ranging from ten tons to eighteen tons or more on one pair of driving-wheels, according to the description of work to be performed.

Very often secondary or branch lines, with comparatively light traffic, have steep gradients, necessitating engines as heavy as on a main trunk line; but the number of trains on the former may not exceed twenty per day, while on the latter they may amount to one hundred and fifty or two hundred. It is evident that the rail which would last for very many years under the small traffic, would have a very short life under the frequent traffic. Hence the reason why it is found expedient to give a large increase of material in the heads of rails carrying the heavy, constant train service of many of our main lines.

Figs. 261, 262, and 263 are sections of rails in use on lines having heavy engines and fast trains, but with a comparatively small daily train service, and Figs. 264, 266, 267, and 268 are sections of rails carrying the heavy, fast, and incessant traffic of some of our leading lines.

On lines having small traffic, slow speeds, easy gradients, and comparatively light engines, a reduced section of rail may be adopted; but in doing so it is well to allow for any probable future development of traffic which might cause the introduction of heavier engines.

Figs. 269 to 272 show sections of rails varying from 72 to 60 lbs. per yard, also a section of a 45-lb. steel flange-rail, much used on 3-foot narrow-gauge railways.

Valuable and interesting statistics have from time to time been recorded, with a view to ascertain the average life of a steel rail, by obtaining the number of million tons of train load which it would sustain before it became worn down to such an extent as to be no longer of service on the line. It will be readily understood that the rate of wear of a steel rail will depend not only on the weight and section of the rail itself, but on the class of rolling-stock, and the description of traffic it has to carry. It will also be largely affected by the circumstances of whether the line is on a level or on an incline.

The writer has had careful measurement taken of the wear of the steel flange-rail (Fig. 265), 79 lbs. per yard, and the result shows that with a traffic not exceeding twenty-four goods and passenger trains per day, one-tenth of an inch was worn off the top of the rail in ten years on the comparatively level portions of the line; but that the same amount of one-tenth of an inch was worn off in six years by the same traffic, on the same district of the line, in places where the gradients varied from 1 in 100 to 1 in 70. The heavy pounding of the engines, and the working of the brakes tend very materially to shorten the life of the rails on the inclines.

As now made, the steel rails manufactured under the converter process exhibit great similarity in the analysis of their component parts; at the same time it is well known that a slight preponderance or reduction of one or more of the constituents will result in making the steel hard or soft. The following statement gives the analysis of twelve steel rails, six of which were classed as *hard* steel, and six as *soft* steel:—

HARD STEEL.—ANALYSIS OF SIX STEEL RAILS WHICH BROKE EITHER IN TESTING OR IN LINE.

	1.	2.	3.
Carbon	0·47	0·51	0·56
Silicon	0·09	0·08	0·08
Sulphur	0·06	0·06	0·06
Phosphorus	0·07	0·06	0·06
Manganese	1·23	1·10	0·90
Iron	98·08	98·19	98·34
	100·00	100·00	100·00

	4.	5.	6.
Carbon	0·43	0·47	0·54
Silicon	0·09	0·095	0·121
Sulphur	0·06	0·054	0·056
Phosphorus	0·08	0·08	0·057
Manganese	1·23	1·15	1·26
Iron	98·11	98·151	97·966
	100·00	100·00	100·00

SOFT STEEL.—ANALYSIS OF SIX STEEL RAILS WHICH STOOD THE TEST WELL, AND BENT FREELY WITHOUT SHOWING ANY SIGN OF FRACTURE.

	1.	2.	3.
Carbon	0·35	0·39	0·37
Silicon	0·06	0·07	0·07
Sulphur	0·062	0·061	0·062
Phosphorus	0·061	0·061	0·061
Manganese	0·870	0·875	0·866
Iron	98·597	98·543	98·571
	100·000	100·000	100·000

	4.	5.	6.
Carbon	0·34	0·35	0·250
Silicon	0·08	0·07	0·069
Sulphur	0·061	0·061	0·046
Phosphorus	0·063	0·062	0·058
Manganese	0·864	0·800	0·636
Iron	98·592	98·657	98·941
	100·000	100·000	100·000

Many rails which have been broken in the line under traffic have been analyzed, and proved to be hard steel; while others, which have been bent into all sorts of shapes, but not broken during accidents or derailments, have also been tested, and proved to be of soft steel.

Some engineers are advocates for a hard steel rail, and claim for it greater durability and longer wear; but even supposing such hard rail should possess a slight superiority over the soft rail, it is well to consider whether such assumed advantage is not obtained at the risk of incurring greater liability to fracture. It must be borne in mind that a rail, once placed in the road, is exposed to all the changes of temperature from heat to frost, and has frequently to sustain increased strains arising from loose sleepers, where the gravel or ballast has been disturbed during heavy rains.

When writing a specification for steel rails, it is usual to state the number of tons per square inch in tensile strain which the steel must be able to sustain without fracture, and also to stipulate that some of the rails will be tested by the falling-weight test. In the latter test a rail is placed, say at 3 feet bearings, and in a similar position to what it would occupy in the road, and a weight of eighteen hundredweight, or one ton or more, according to section of rail, is allowed to fall from a height of 9 or 10 feet, on to the rail, at the centre between the bearings. With three blows from the given height, the rail must not bend or deflect more than a specified amount. The falling-weight test is, perhaps, rather a rough and ready one; but it is always reassuring to prove that the rails will withstand such a severe ordeal, as it must be a very exceptional circumstance in the routine of railway working which will produce a blow or shock equal in effect to the falling-weight test. The rails form such an important part of the trackway, almost the very basis on which the traffic has to depend for its safety, that, apart from the question of wear, no effort should be spared to ensure their thorough soundness and efficiency.

In modern practice rails are generally used in lengths varying from 25 feet to 30 feet. There is no difficulty in making them longer; but any excess over the above lengths is found to be inconvenient for transport, for handling in the line, and for making the necessary allowance for contraction and expansion at the joints. Steel rails are generally marked on the vertical web with the initials of the railway company, the name of the manufacturer, and the year in which they are rolled. This is done by cutting out the letters in the last pair of rolls through which the rails have to pass before they are completed, so that on the rails themselves the letters stand out in raised characters, thus: **G.N.R.I.......C. CAMMELL & Co 1896**. In this manner the rails always carry for reference the name of maker and date.

When comparing the relative merits of the flange-rail and bull-head-rail permanent way, the question of strength and durability must be considered, as well as that of economy. The flange-rail road has undoubtedly fewer parts and fastenings, and when the flange is wide, the sleepers sound, and the rail securely held down to the sleepers, the result is a smooth running road. So long as the rail can be maintained in a constant close contact with the wooden sleeper, the running is almost noiseless, the jarring on the rails being absorbed or taken off by the timber; but so soon as a little space or play takes place between the spikes or other fastenings and the upper surface of the flange, the rail obtains a certain amount of rise, or lift, which comes into action upon the passing of every rolling load, producing unsteadiness in the rail and a clattering noise in the running. A flange of 5 inches, on a sleeper 10 inches wide, has a bearing surface of 50 square inches (assuming the sleeper to be square cut, without any wane on the edges), and this area of 50 inches is only about half of the bearing surface on the sleeper of an ordinary modern cast-iron chair.

Main-line locomotives have weights on the driving-wheels varying from 16 to 18 and 20 tons. Taking 18 tons as representing a common practice for a large express engine, would give 9 tons as the weight imposed on each rail by each driving-wheel Assuming this weight to be distributed over three sleepers would give a dead weight of 3 tons per sleeper, or 134 lbs. on every square inch of the 50 square inches of surface, or rail-bearing area, on each sleeper, without taking into account the effect of the blow or percussion from the rolling load. The presence of a loose sleeper throws additional weight on the adjoining sleepers, and increases the destructive influence on the timber. The constant application of heavy rolling loads on a small bearing area of timber crushes and wears away the timber very rapidly. The small bearing surface of the flange rail expedites the cutting down into the sleeper, and as the rail beds itself further and further into the wood, the fastenings must be driven or screwed down to follow the flange. Spikes may be driven down, but the further they go they have a less thickness of timber for a bed, and therefore a diminished hold. Crab bolts are apt to become rusted or ironbound, so that they cannot be screwed further, and must then be taken out and replaced with new ones. The narrower the flange, the more rapidly does the rail-seat cut down to a thickness inconsistent with safety. The sharp edge of the flange-rail has a tendency to cut a channel in the spike, and it is not at all an unusual occurrence to find strong square shanked dog-spikes, which have been thus cut into to the extent of a third or even half their thickness. The comparative narrow flange places the spikes at great disadvantage in point of leverage for holding down, and this weakness is soon made manifest, particularly on curves, where additional crab bolts or other devices are rendered necessary to counteract the tendency of the rail to rock and tilt over sideways. When the head of the rail cannot be kept in its proper position, the gauge becomes widened, and an irregular sinuous motion takes place in the running of the train. This drawback has been found to be a serious matter where light narrow flange rails have been adopted to carry comparatively heavy, short wheel-base engines. In some cases wrought-iron sole-plates, or even cast-iron bracket-chairs, have been introduced to give more bearing surface on the sleeper and increased support to the rail, but neither of the two methods give the same simple complete hold to the rail that is obtained by the cast-iron chair for the bull-head rail.

On the other hand, the modern cast-iron chair for the bull-head rail has at least double the bearing surface on the sleeper to that of the flange-rail seat, so that under the same circumstances of rolling load as above described, the weight of 134 lbs. per square inch would be reduced to half, or 67 lbs. The greater length given to the chair effectually prevents any rocking action on the part of the rail, and reduces to a minimum any lifting action on the spike. A good fitting chair—especially when keyed on the inside—provides a most effectual support to the rail both vertically and laterally, and maintains the rail to accurate gauge. By giving proper clearance space at the tops of the chair-jaws, a bull-head rail can be taken out by simply driving out the wooden keys, and a new rail inserted without in any way disturbing the chairs or spikes. To change a flange rail necessitates the slackening and removal of a large number of the spikes and crab bolts.

As the sleepers under the chair road suffer less from the crushing of the timber, they have a much longer life in the line, and remain serviceable until they are incapacitated from decay. This is a very important item in places where timber sleepers are expensive. The steadiness of the chair prolongs the efficiency of the spikes.

As the actual wearing portion of the rail is the head, or wheel contact surface, a liberal area—consistent with the expected traffic—must be given to that part, whether for a bull-head rail or a flange rail. By comparing the two sections, Figs. 273 and 274, the one for an 85-lb. bull-head rail, and the other for a 100-lb. flange rail, it will be seen from the dotted lines that the heads of each rail are almost identical, the difference of 15 lbs. being disposed of in the flange of the heavier rail. Practically, therefore, we have 15 lbs. per yard extra weight of steel in the rail, on the one hand, as against the cast-iron chairs and steadier permanent way on the other.

For lines where the traffic is small, weights light, speeds low, and economy of construction imperative, the flange-rail permanent way will be very suitable.

The writer has had long mileages of each description of permanent way under his charge, both at home and abroad, for many years, and the result of his experience has shown that, although a fairly good road may be made with flange rails, still, for constant, heavy, fast traffic, the bull-head rail with cast-iron chairs makes a much stronger, more durable, and better permanent way than any flange railroad.

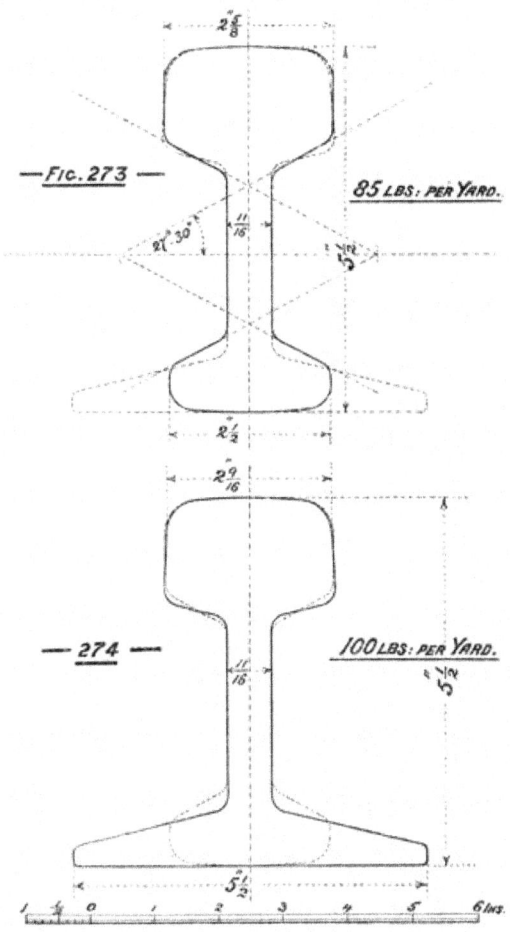

Briefly summarized, the principal advantages and disadvantages of the two kinds of rails stand as follows:—

ADVANTAGES.

Bull-head Rail.	Flange Rail.
Large bearing surface of chair upon the sleeper, and	Fewness of parts, and less cost.
	Smaller quantity of ballast required

Bull-head Rail	Flange Rail
greater stability of the rail.	to cover up the foot of rail.
Longer life of wooden sleeper.	More lateral stiffness than the bull-head rail.
Impossibility of rail tilting over outwards.	
Facility for changing a rail without disturbing the fastenings in the sleepers.	
Easier to maintain, owing to less disturbing strains on the fastenings.	
A bull-head rail is more readily set or laid to follow line of curve.	
In most cases the one set of chairs will serve for a second set of rails.	
Perfect straightness of rail: it is very rare to find a crooked bull-head rail.	
Easier to roll, and more likely to obtain uniformity of steel.	

DISADVANTAGES.

Bull-head Rail.	Flange Rail.
Greater cost.	The small rail-seat area on sleeper throws great crushing weight on the timber.
More ballast required to	

cover up the rail.	Shorter life of wooden sleepers from the cutting down of rail-seats.
Less lateral stiffness than the flange rail.	
	The edge of flange cuts the spikes after a few years.
	The undulation of the rail under trains tends to raise the spikes, and causes lateral movement in the rails.
	More difficult to maintain, in consequence of greater tendency of the fastenings to work loose.
	Difficulty in getting flange rails straightened laterally.
	More difficult to set to follow regular line of curves.
	More difficult to roll, and less likely to obtain uniformity of steel.

Tramway Rails.—Tramways on streets or public roads are now universally recognized as important branches of the railway principle. Their smoothness of movement, increased accommodation, and many other advantages as compared with the old road omnibus, render it no longer necessary to call for special advocacy when there is a possibility of their introduction. They occupy a position so thoroughly appreciated by the public that any check on their reasonable use or extension would be considered as detrimental to the interests of the travelling community.

As a rule, these tramways are laid down on streets or roads previously constructed for the ordinary road traffic, where all the preliminary work of earth filling, bridges, drainage, etc., has already been accomplished, and there only remains the selection and laying down of the rails or permanent way over which the tram-cars will have to run. The description and weight of permanent way to be adopted will depend largely upon the weight of the cars to be used and the system of motive-power decided upon for the haulage—whether horses, steam, cable, or electricity.

As the portion of the streets or public roads along which the tramway has to be laid will, in all probability, have to be occupied and traversed by all kinds of vehicles besides the tram-cars, it is absolutely necessary that the permanent way for the tramway should be of such description as to require the least possible amount of adjustment of fastenings or opening out of the roadway for repairs. Where the entire width of the street, including the space between the tram-rails, is paved with stone setts, the opening out of even a short length for repairs is tedious and costly, and causes considerable obstruction to the street traffic. It is most important, therefore, that the rail and its fastenings should not only be strong enough for its own tram service and the carts and drays which will pass over and across the track in all directions, but it must possess the minimum necessity for disturbance.

Figs. 275 to 279 are sketches of a few of the many types which have been brought into use in various places.

Where the public roads are wide, and a space can be set apart at the side for the special use of the tramway, the arrangement shown in Fig. 275 will be simple and efficient. It is very similar to an ordinary railway permanent way with the ballast filled in flush with the top of the rails. The rails are shown as flange or flat-bottom rails, fished together at the joints, and properly secured to transverse sleepers of wood, iron, or steel. The space between and outside the rails is filled in with small-sized broken stone ballast or good clean gravel, and forms an even surface, over which animals or cattle may pass without risk of being thrown down.

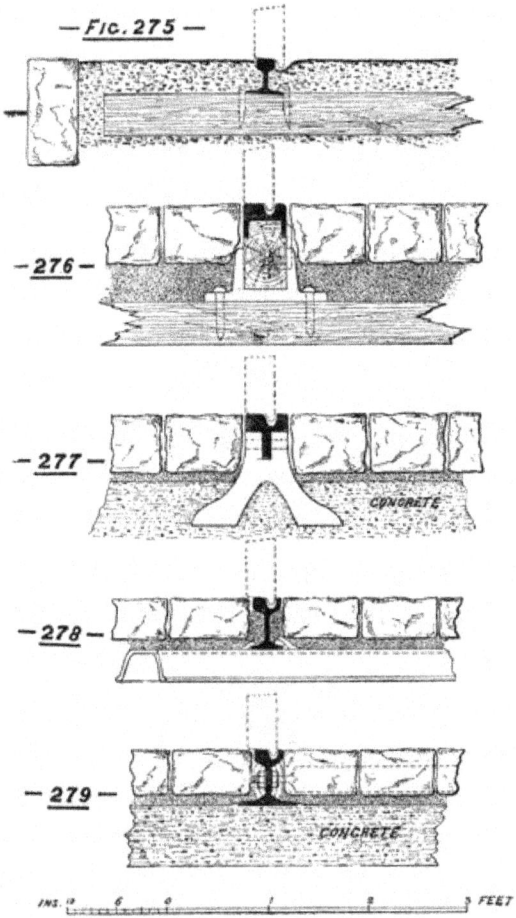

Fig. 276 represents a system which was laid down extensively, especially for horse tramways, but not proving efficient, has been superseded by other types of a stronger and more durable description. The rail was rolled with a continuous groove to provide clearance for the flanges of the car-wheels, and the sides of the rail were turned down so as to fit over the longitudinal timber sleeper, to which the rail was secured by staple-dogs, as shown. Cast-iron chairs, spiked on to wooden cross-sleepers, held the longitudinal sleepers in position. The wooden sleepers were favourable for smooth running, but the section of the rail, practically a light channel-iron laid on the flat, was most unsuitable for carrying weight or for making a proper joint. Experience proved this road to be very difficult to maintain in good order for easy traction. The staple-dogs worked loose after a little time, and the rail, having scarcely any vertical stiffness, rose and fell during the passage of every car-wheel, resulting in most uneven joints and a clattering roadway.

With the view to obtain a stronger and more permanent support for the rail than the longitudinal timber sleeper last described, various forms of cast-iron chairs were devised. Fig. 277 represents one of these patterns. The rail, which is of **T**-section with a continuous wheel-flange groove, is secured to the cast-iron chair by the cross-pin, as shown. Although this cross-pin may in time work a little loose, it cannot work out, being kept in position by the paving-setts on each side. The cast-iron chairs are placed at convenient distances, and being set in a bed of concrete, do not require cross-sleepers or tie-bars. This type makes a strong road, but the rail-joints cannot be made so even or efficient as with the more modern form of rail.

Rail manufacturers are now able to roll a section of rail combining the vertical stiffness of the ordinary flange, or flat-bottom, rail with the running-head and continuous wheel-flange groove, considered the most suitable for heavy tramway traffic. The introduction of this section of rail has contributed greatly to the increased efficiency and durability of the permanent way for street traffic; and as the ends of the rails can be secured by ordinary fish-plates, there is the great acquisition of even joints and increased smoothness in the running of the tramcars. This rail can be rolled of various weights to suit the rolling loads. On some tram-lines a moderately heavy section has been adopted, and secured to transverse sleepers of rolled iron or steel laid on a bed of concrete. On others similar rolled metal sleepers have been used, but laid longitudinally. For some descriptions of traffic a much heavier section of rail has been used, having a base sufficiently wide to provide ample bearing on a bed of concrete without the intervention of either transverse or longitudinal sleepers.

Fig. 278 is a sketch of the modern rail as laid down on a rolled steel transverse sleeper, the rail being held in position either by turned-up clips, wedges, bolts, or any of the devices in use for similar duty in the rolled-steel sleepers for ordinary railway permanent way.

Fig. 279 shows a modern rail of a heavier section, with a wide flange resting direct on a continuous bed of concrete. The gauge is maintained by bar-iron tie-bars placed vertically so as to fit in between the courses of the paving-setts, the ends being forged and screwed to pass through holes in the vertical web of rail, and secured in position by nuts. Both in this, and in type Fig. 278, ordinary fish-plates are adopted at the rail-joints, as indicated by dotted lines.

In the last two examples above described all the materials are of the most durable description, and the least liable to wear or decay, but it will be necessary to guard against making the fastenings and the bars too light for the duty they have to perform. There should be ample material in the head of the rail to allow of a fair wearing down, and the continuous flange groove should be sufficiently deep to meet this wearing away without causing the wheel-flanges to strike the bottom of the groove.

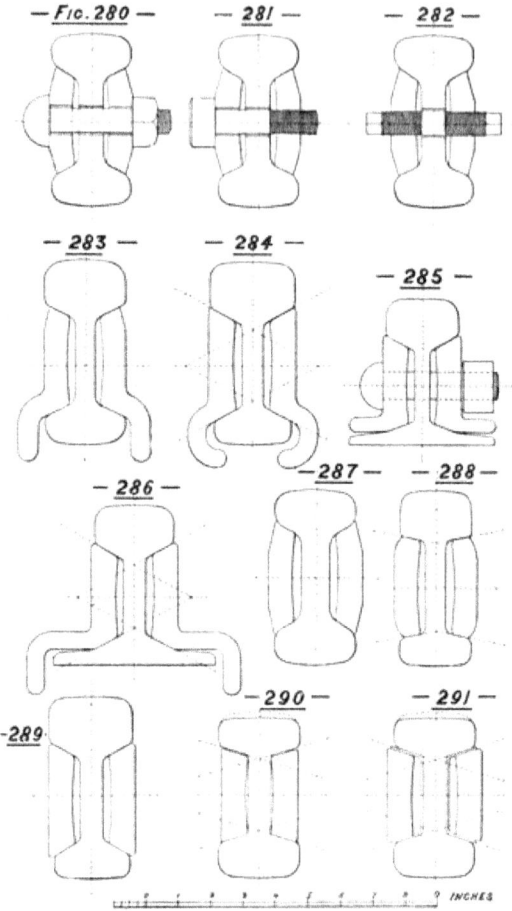

Fish-plates.—In the first examples of the newly invented wrought-iron fish-plates they were made to the depth to fit in between the upper and lower tables of the rail, as shown in Fig. 280, a small space or clearance being left between the inner sides and the vertical web of the rail. Ordinary nuts and bolts were used in most cases, but in some instances one of the fish-plates was tapped, as in Fig. 281, forming one long continuous nut, and in others both fish-plates were tapped, as in Fig. 282, and right and left handed bolts were used. Neither of the two arrangements of tapped fish-plates proved sufficiently successful as to lead to their general adoption. When the bolts became rusted in, or iron-bound, it was found to be almost impossible to remove them without permanently damaging the fish-plates. With the four right and left handed bolts the operation of tightening, or removing, the fish-plates was very tedious, as each bolt had to be turned a very little at a time, one after the other. Independent bolts and nuts, either of iron or steel, are now universally used; plain holes, with sufficient allowance for work and expansion, being punched or drilled in the rails and fish-plates.

For many years the depth of the fish-plates continued to be made the same as the space between the upper and lower members of the rail, as shown in Fig. 280; but with the heavier loads and higher speeds of our modern railway working it has been found necessary to strengthen the joints by providing deeper or stiffer fish-plates, as shown in Figs. 283, 284, and 285. For bull-head rails the fish-plates have been brought down underneath the lower table, and in some cases extended down sufficiently far to admit of a second set of fish-bolts under the rail. For flange rails some fish-plates are used simply of the form of angle irons, and others have the angle portion carried out beyond the end of the flange, or foot of rail, and then turned down vertically to a depth of an inch or more below the rail. The latter makes a very strong fish-plate.

Fish-plates, like rails, are now almost universally made of steel.

The efficiency and durability of a fish-plate depends materially upon its angle of contact with the under side of the head of the rail, and the extent of its contact surface. It would be an error to suppose there is little or no wearing away in fish-plates, as in reality there is very considerable wear, and especially in rails of lighter section. If the under side of the head of rail has a curved outline, as in the rail in Fig. 287, there will be some difficulty in ensuring a perfect fit in the fish-plates; the curve of the one may not quite correspond to the curve of the other, and the contact surface will be very small. It is better to make these contact surfaces in straight lines, and to a wide angle rather than to an acute angle. In Fig. 288 the under side of head and corresponding top of fish-plates are set at an acute angle, and fish-plates to this pattern will soon wear up to the vertical web of rail, and cause a loose noisy joint.

In Fig. 284, showing a different type of rail, the contact surfaces are set at a very much wider angle, and will allow much more wear before the fish-plates can work close up to the web of the rail.

When once the fish-plates are close up to the web, the best and tightest bolts cannot prevent the vertical play in the ends of the rails.

A hammering sound will announce each successive drop of the wheels from one rail to the other, more distinctly, perhaps, at slow speeds than when travelling quickly, but existing equally under both conditions. The unpleasant jarring sensation is annoying to the passengers, and has a straining, loosening effect on all the bolts and fastenings. Unless the fish-plates have a thorough continuous bearing against the upper and lower shoulders of both the rails, it will be impossible to obtain a smooth even joint. A road may have good rails, good chairs, and good sleepers, but if the fish-plates are worn and loose the entire permanent way may be pronounced faulty, and all on account of a minor defect which can be easily remedied. With strong, properly fitting fish-plates, the position of the joints should be imperceptible when passing over them in a train.

The writer has had many miles of line where the fish-plates have worn hard up to the rail web. In cases where the rails were good, with the prospect of a long life, new fish-plates of suitable section have been provided. In others, thin wrought-iron plate liners, 1/16 or 1/12 of an inch thick, have been inserted, as in Fig. 291, so as to bring the plates well out from the web, and allow the fish-bolts and fish-plates to exercise the free gripping action which is absolutely necessary to prevent the vertical rising and falling of the rail-ends during the passage of a rolling load. Fish-plate liners of the above description have given excellent results, and have restored the efficiency of the fish-plates for several years.

Chairs.—All rails which partake of the double head section, or have a base not wider than the head, require supports or carriers to attach them to the sleepers, and to secure them in their proper upright position. In the days of the original *edge rails*, at the commencement of the railway era, these supports were very appropriately termed *chairs*, and this name has now been adopted in all parts of the world. Cast-iron is the most suitable material for railways chairs, being much cheaper in cost and less liable to loss or deterioration from rust than wrought-iron. Cast-iron chairs can be formed to suit any section of rail, and from the nature of the material they cannot be bent or twisted out of shape so as to interfere with the gauge or cant. They may break during an accident or derailment, but the fracture can be detected at once, and the broken chair quickly replaced.

The chair performs the very important duty of distributing the weight of the rolling load on the upper surface of the sleeper. If the under side or base of the chair is small, and the rolling load large, the chair will very rapidly wear or imbed itself into the wood of the sleeper, shortening the life of the latter in a very palpable manner. The short narrow chair naturally gives less stability than the larger and broader chair. The chair shown in Fig. 292, which was much used for 75 lb. rails some twenty years ago, has much less base area and stability than the chair shown in Fig. 293, adopted for rails of a similar weight in the present day. The former had a bearing surface on the sleeper of only 53 square inches, as compared with 89 square inches in the latter. The base area of the chair must be in proportion to the weight it has to carry and distribute, and it would be false economy to stint the surface area of one of the details which influences so materially the stability and durability of the permanent way.

As will be seen in Figs. 294, 295, and 296, the chairs at present used for 80, 85, and 90 lb. rails have a much larger bearing surface than the chair shown in Fig. 292.

With the wider chair, a much longer and better seat can be given to the under table of rail, and a greater length of jaw for holding the wooden key. The longer the rail-seat the steadier the rail and the smoother the running.

The keys are generally made of hard wood, sometimes compressed by a special process, cut slightly taper, or wedge, shape, and driven in between the jaw of the chair and the vertical web of the rail. On some railways the key is placed outside the rail, as in Fig. 297, and on others inside the rail, as in Fig. 298. The latter method possesses many advantages over the former. The outer jaw of the chair can be brought well up to the under side of head of rail, giving the rail more lateral support and better means of preserving the correct cant; and, as in this chair the outer jaw permanently fixes the gauge, the working out of one or more of the keys does not leave the rail exposed to be forced outwards and widen the gauge, as in the case with dropped keys in outside keying. Another and very important advantage of inside keying is that platelayers, when inspecting the road by walking between the rails, can readily examine the keys on both sides.

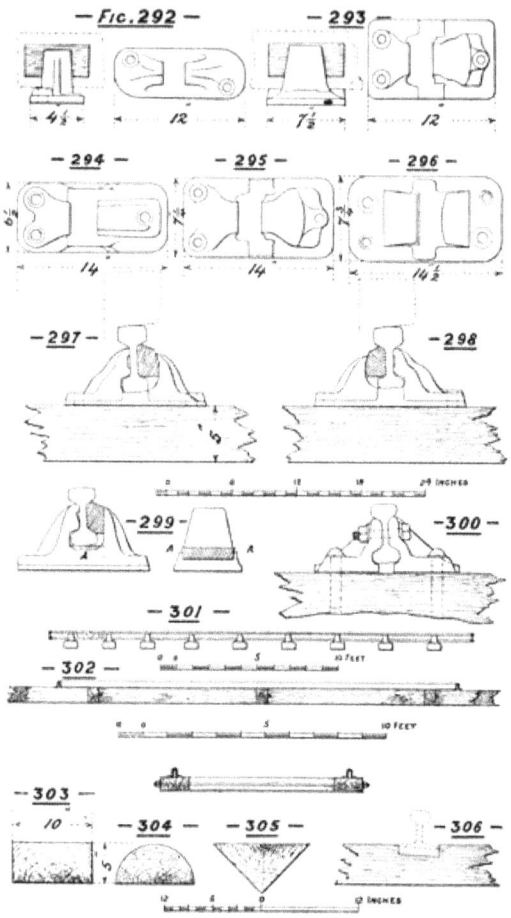

Chairs have been made, as in Fig. 299, with a recess in the rail-seat, to hold a piece of prepared wood, or other suitable semi-elastic material, the object being to provide a rest, or cushion, softer and more yielding than the cast-iron. The idea looks well in theory, but in practice the pounding on the rail compresses or crushes the wood lower and lower into the recess, slackened keys have to be tightened, and when the wood has been worn or crushed away down to the level of the stop ribs, **A**, **A**, the under side of rail has no longer any seat, or rest, beyond the two narrow ribs of cast-iron. These afford such a very limited support that the rail becomes notched, and produces a very rough clattering road. It is a very simple matter to take out an old key and put in a new one, but to replace a wooden cushion in a chair recess involves the entire removal of either the rail or the chair. Chairs with wooden cushions have not been adopted to any great extent, the tendency of modern practice being to reduce as far as possible the number of parts of the permanent way, and to provide those parts with ample bearing or contact surfaces.

Although the general practice has been to cast the chairs in one piece, chairs have been made in two pieces, as in Fig. 300, fastened together and to the rail by a bolt passing through the latter, the castings being secured to the sleeper with spikes. At first sight this pattern of chair appeared to possess some features in its favour. The castings were simple, keys were dispensed with altogether, and the under side of rail was not in contact with the cast-iron. A short experience, however, proved that the drawbacks far outweighed the apparent advantages. Holes for the through-bolts had to be punched at fixed distances in the rails, and although this could be readily done at the works, for the general use on the line it was necessary to resort to the tedious process of drilling by hand for a large number of holes on curves, and for rails cut to form *closers*.

Sleepers.—Wood possesses so many suitable qualities that we can readily understand why it was early selected as the proper material for sleepers. It can be cut to any size and shape, holes can be bored, spikes can be driven, and bolts can be screwed into it without any difficulty and without causing injury to the timber, while the semi-elastic nature of wood absorbs the vibration of the rails and fastenings, and provides a sound-deadening seat so conducive to smooth running. Its only drawback is that it is perishable from wear and decay. Were it not for this defect, railway sleepers of wood might be considered as simply perfect.

With a view to greater permanency and durability, stone sleepers were tried. These consisted of square blocks of good hard stone, measuring about 2 feet wide each way and 12 inches thick. Holes were cut in the stone, and plugs of hard wood inserted. The cast-iron chairs were then placed on the top of the blocks, and the iron spikes driven through the chair-holes into the wooden plugs. The elements of permanency were there certainly, but a rougher road it would be impossible to conceive. The stone was solid and unyielding, there was a total absence of softness and elasticity, and the harsh noisy effect produced when running over the stone-block road very soon became intolerable. Stone-block sleepers were found to be a failure, and were all removed. On some of our old lines, numbers of them, with the chair marks plainly visible, may be still seen in loading banks, buildings, sea walls, and other works for which they were never originally intended, but for which their size and weight render them very appropriate.

Wooden sleepers are used in two forms, transverse and longitudinal. In the former, as in Fig. 301, the sleepers not only carry the rails, but also preserve the gauge; in the latter as in Fig. 302, the longitudinal sleepers only support the rails, additional timbers and strong fastenings being necessary to maintain the gauge.

Longitudinal sleepers have been used to a large extent for bridge rails, it being supposed that with the broad continuous sleeper a lighter and shallower rail could be adopted, which would be equally efficient as a heavier rail on cross-sleepers. Excellent running roads have been made with longitudinal sleepers, notwithstanding the difficulty of making a good bridge-rail joint; but it is well to bear in mind that almost all the lines which originally adopted this form of permanent way have since reverted to the ordinary cross-sleeper road. The longitudinal sleeper road is an expensive road to lay down and maintain. The main pieces are of large scantling, must be of good quality of timber, and are consequently costly. The cross-pieces, or transomes, must be carefully fitted and secured with heavy ironwork. Where there is much traffic, the removal and renewal of one of the long timbers is much more difficult than the renewal of several ordinary cross-sleepers. Again, decay may take place on only one portion of a main timber, but there is no alternative but to remove the entire piece.

For gauges varying from 4 feet 8½ inches to 5 feet 3 inches, cross-sleepers are cut to the length of 8 feet 11 inches, and are generally rectangular in section, as in Fig. 303, measuring 10 inches in width by 5 inches in thickness. On some of the lighter railways with small traffic, sleepers are often used only 9 inches wide by 4½ inches thick, while occasionally on some lines, and in places where there is exceptionally heavy and constant traffic, sleepers 12 inches wide by 6 inches thick are adopted.

Half-round sleepers, as in Fig. 304, are used on many lines because they are cheaper. In some cases the flat side of the sleeper is placed downwards, and the rail or chair is fastened into an adzed seat cut in the round side; and in the others the round side is placed downwards, and the flat side of the sleeper carries the rail or chair. Triangular sleepers, as in Fig. 305, have also been used, made by cutting the blocks diagonally, so as to obtain the greatest possible width. They were laid with the flat side upwards, and the apex downwards. They were difficult to keep packed, and have not been adopted to any great extent.

With the exception of a limited number of larch and fir sleepers grown in the country, most of the sleepers for our home railways are imported from the Baltic. They are brought over in logs, or blocks, each 8 feet 11 inches long, some square and others circular in section, and when sawn down the middle, each block forms two sleepers.

The preservation of timber from decay is a subject that very early occupied the attention of engineers and all those interested in railways. A railway sleeper is particularly exposed to deterioration the lower portion being surrounded with moist ballast, whilst the top portion is more or less uncovered—two different conditions in the same piece of timber. Several processes have been tried, such as Kyanizing, Burnetizing, Boucherizing, etc., but the system which has given the best results, and is now almost universally adopted, is that known as creosoting. This method consists of forcing liquid creosote, under considerable pressure, into sleepers or railway timbers which have been prepared or dried by ordinary natural seasoning or by special artificial means. Creosote is a dark, oily liquid, distilled from coal tar, varying in its composition according to the quality of the coal from which it is obtained, and ranging in its specific gravity from 11·08 to 10·28.

Creosote oils of light specific gravity were at one time in favour, but experience proved that, to some extent, the light oils were volatile and also soluble in water, and that heavy rains washed out the constituents which were essential for the preservation of the timber. On the other hand, by heating the heavy oils and using high pressure the napthaline which is dissolved only by the heat, is forced into the wood, fills the pores, and solidifies.

Creosote is obtainable in large quantities, at prices varying from twopence to fourpence per gallon, according to the demand and cost of production. Newly delivered sleepers or railway timber contain so much sap or water that it is impossible to force a sufficient quantity of creosote into them until they are properly seasoned or dried.

The seasoning is generally arranged by sawing each block into two sleepers, and then stacking the sleepers on edge in tiers, leaving a space of four or five inches between each of them for a proper circulation of air. The sleepers should then be left for nine to twelve months to season, although more may be necessary in some cases if the blocks were particularly wet at the time they were sawn.

When ready for the process the sleepers are placed in the creosoting cylinder, which is generally about 60 feet long by 6 feet in diameter with semi-spherical ends. One of the ends is fitted with strong hinges and fastenings, and forms the doorway. The sleepers are packed carefully inside, and the doorway made tight. The machinery is then set to work to exhaust the air from the cylinder and allow the creosote to flow in amongst the sleepers. When the cylinder is full the force-pumps are started to force in more creosote up to the pressure prearranged and regulated by the safety-valve, in some cases 100, 110, or 120 lbs. per square inch. The creosote should be heated to 112° or 120° Fah., to dissolve the napthaline and reduce all the component parts to a thoroughly fluid condition.

The success of creosoting depends almost entirely upon the effectual seasoning of the timber. Only a very small quantity of creosote can be forced into wet or unseasoned sleepers, even with the best machinery and exceptionally high pressures, while a thoroughly dry sleeper will readily absorb from $2\frac{1}{3}$ to 3 gallons. More could be forced into the dry sleeper if necessary, but a little consideration will show there would be no advantage in doing so. In railway sleepers there are two elements of destruction at work—one the decay of the timber, and the other abrasion or wearing away of the wood itself from the constant pounding of the passing loads.

More particularly does this wearing-away take place with the flange, or bridge, rails, their distributed bearing surface on the sleeper being less than the cast-iron chairs.

A thoroughly well-creosoted 5-inch sleeper laid originally with a thickness of 4-¾ inches in the centre of rail-seat, as in Fig. 306, will wear down 1½ inches, the timber remaining quite sound.

The writer has had to take out thousands of sleepers where the seats of the flange, or bridge, rails had been pounded or worn down so deep into the wood as to leave too small a thickness of timber to carry the rail with safety. These sleepers had to be taken out of the road, not on account of decay, but because they were actually worn down too thin to be of service. They had done their work well for a long series of years, and were perfectly sound when taken out. No increased quantity of creosote would have made them last longer, and any increased quantity of creosote would have been waste.

Two and three quarter gallons of creosote is a very good and suitable quantity for a 10 inch by 5 inch rectangular sleeper, but not more than half this quantity can be forced in if the sleeper is wet or unseasoned.

Sleeper-blocks are generally cut from the upper part of the tree, and do not therefore consist of the best portion of the timber, yet sleepers made from the soft, coarse-grained Baltic wood, properly creosoted, will last from twelve to eighteen years in the line in this country, while uncreosoted they would perish from decay in six or seven. The benefit is great when, by adding from eightpence to a shilling for the cost of creosoting, the life of the sleeper may be doubled or trebled. Of course, there are countries, like the far west of America, where the lines pass through vast forests, and where sleepers may be had for the mere cost of cutting. Creosoting in those places would be out of the question, and would cost four or five times the value of the plain sleeper. It is found, also, that in tropical countries and in dry climates at high altitudes creosote loses its efficiency, and in those districts the best creosoted soft-wood sleeper perishes from a species of dry rot in three or four years. Where wood sleepers have to be used in tropical climates it is better to obtain them from the timber of the district, although in many cases suitable trees are difficult to procure and the cost of land transport is very heavy.

The soft cushion-like effect of a sound, properly packed wooden sleeper contributes so largely to form an easy, smooth-running road, that so long as they can be obtained at a moderate cost, and are fairly durable, wooden sleepers will always be preferred to those of any other material. The great question will be the supply. Creosoting and other wood-preserving processes have done much to prolong the life of sleepers, but the rapidly increasing extent of mileage throughout the world, together with the enormous number of sleepers required annually for maintenance or renewals, must before very long severely tax the powers of supply.

In the great timber-producing territories the axe is often heard, but the planter is rarely seen. Vast forests are cleared away, and their sites transformed into busy towns or cultivated lands; and unless some great change takes place, and planting be carried out on a large scale, some other material will have to be adopted for this important item of our permanent way.

Appearances would indicate that at no very distant date iron or steel will take a conspicuous part in the formation of future railway sleepers.

More than thirty years ago several descriptions of cast-iron sleepers were introduced into notice and tried on some of our leading home railways. Cast-iron was at that time considered more suitable for the purpose than wrought iron, as it was very much less costly in price, and could be readily worked into any desired form or size, with the advantage that the castings would all be duplicates of one another.

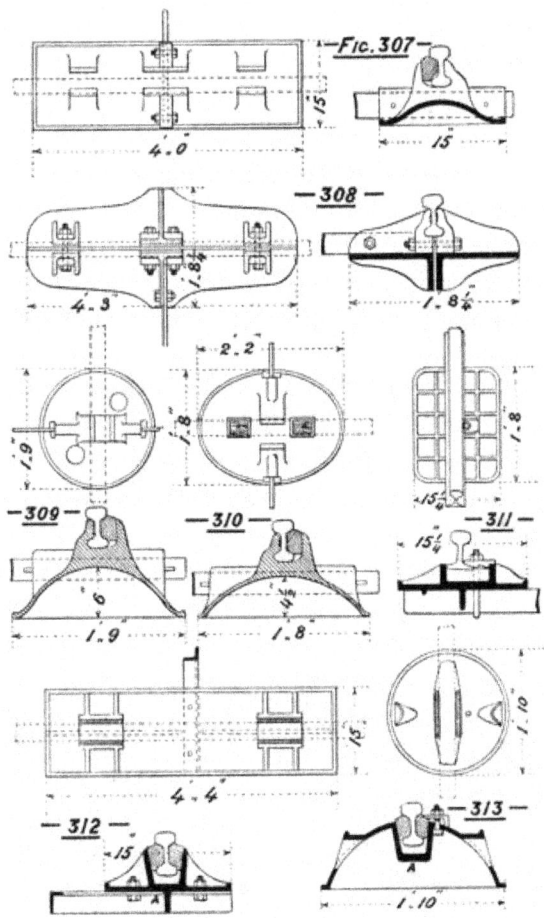

Figs. 307 to 313 show some of the types that were designed and laid down in the road. In Fig. 307 the sleeper and chairs were all cast together in one piece; the rail was held in its place by wooden keys, and the gauge of the line was maintained by transverse wrought-iron tie-bars. The sketch represents one of the sleepers used at the rail-joints, and has three chairs, the larger one in the centre being for the support of the ends of the rails. This arrangement was the same as was then in use on the ordinary wood-sleeper road, where an extra large chair was placed at the rail-joints, and was the most approved method for many years before fish-plates were introduced. The intermediate sleepers were shorter, and had only two chairs.

Fig. 308 represents a long, flat, cast-iron sleeper made in two halves, bolted together just below the under side of rail at each of the three chair-seats. The rail was gripped and held in position without the use of wooden keys. This being a joint sleeper, three chairs were used, as in Fig. 307. Only two chairs were used on the intermediate sleepers.

Figs. 309 and 310 are somewhat similar, but the circular one is higher and more cup-shaped than the other of oval form. The oval pattern has two small recesses for holding two small hard-wood cushions. The circular holes shown in the sides of the sleepers were intended to facilitate the packing, or tamping, of the light sandy ballast.

Fig. 311 represents a rectangular cast-iron sleeper, as used for the flange rail. The rail rests on cast-iron cross-ribs, bevelled to give the proper cant, and is held in position by the tie-bar bolt and clip-piece, as shown. The small projecting lug, formed on the under side of sleeper, fits into a corresponding notch in the tie-bar, and keeps the sleepers to gauge. The tie-bar passes through the loop end of the same bolt which secures the rail, and is held up tight against the under side of sleeper.

Figs. 312 and 313, both the same in principle, possessed features which appeared to give great promise. They were simple in construction; the rail was kept well down, and did not come in contact with the cast-iron at any point. The long wooden wedges, which fitted into the rough or serrated sides of the casting, acted as a cushion to the rail, and were intended to sink deeper into the recess as the super-imposed weight increased, or the wood became thinner from shrinkage. In practice, however, it was found that these sleepers were not the success that was anticipated.

It was soon observed that sand and fine particles of gravel from the ballast worked their way into the lower part of the recess, and became so compact as to prevent the wooden wedges working further down to increase their grip on the rail. Even when the recess was kept free and clear of sand, the enormous pressure exerted by the wooden wedges broke the iron at **A**, although an extra thickness was given to that part of the section. The cast-iron was exposed to the greatest strain at the point where it was the least capable of offering resistance.

Much ingenuity was displayed in many of the patterns brought forward, but in dealing with a hard unyielding material like cast-iron, it is difficult, if not impossible, to impart any soft, elastic effect; and the different systems of cast-iron sleepers failed to become popular on our home railways, on account of the noise and vibration when trains passed over them. Another objection was the great multiplicity of parts required in many of the types, and the constant and severe strain produced on the fastenings on the passing of every wheel. The bolts might be made tight at first, but the incessant shaking would work them loose, the threads became stripped, and the rails ceased to be held in a proper and secure position.

The cast-iron sleeper road was considered unsuitable for the heavy and fast traffic of our home lines, and was ultimately all taken up and replaced with wooden transverse sleepers. At the same time, there is no doubt that cast-iron sleepers have been of great value in India and tropical climates, where timber sleepers were not only scarce, but perish very rapidly. Very large numbers of them have been laid down abroad of patterns very similar to those shown in Figs. 309, 310, and 311, and have done good service for many years. They are not affected by rain or heat, but, unfortunately, being castings, are liable to considerable annual loss from breakage.

Improvements in plate-rolling machinery, and in appliances for bending and stamping wrought-iron, have materially assisted in developing the introduction of wrought-iron and steel sleepers. Cast-iron and wrought-iron are, in the abstract, hard and non-elastic as compared with wood; but whereas cast-iron can only be made into fixed, unyielding shapes, wrought-iron and steel can be worked into forms that possess a certain spring-like effect, which not only enables them almost entirely to resist fracture, but also imparts a measure of elasticity to the permanent way.

The simplest form of wrought-iron sleeper would be a plain, flat plate, to which the chair, or rail-bracket, would be attached; but as this form would have bearing surface only, without any lateral hold on the ballast to keep the rails to line, it could not be adopted.

During the last few years very many types of wrought-iron and steel sleepers have been introduced, and nearly all of them of the transverse-sleeper pattern, formed out of rolled plates; the sides, and in some cases the ends also, are bent, or turned down to obtain a hold in the ballast. Where bull-head or double-head rails are used, cast-iron chairs, or wrought-iron bracket chairs, are bolted, or otherwise secured to the upper surface of the sleeper, a layer of felt, tarred paper, or other soft material being placed between the two metal surfaces. Where flange rails are used, they are fastened to the sleepers either by bolts, clamps, or clips raised up out of the iron sleeper, and bent over to hold tightening keys. Rolled transverse sleepers can readily be bent, or set in the centre to give the proper cant at the rail-seat; and in some types the sleepers are pressed in the machines, so as to be narrower towards the centre, and with a deeper turnover, to obtain increased stiffness.

In Figs. 314 to 319 are shown some of the patterns which have been brought out, laid down in actual practice, and in use at the present time.

From the fact that wrought-iron and steel sleepers have been laid down in so many places where cast-iron sleepers were discarded or refused a trial, it is evident that the former are considered to have qualities which the latter did not possess. Rolled iron or steel sleepers are coming more and more into use, especially on foreign or colonial railways. So long, however, as good, well-creosoted timber sleepers can be obtained for our home railways at prices from 3*s*. 8*d*. to 4*s*. 8*d*. each, and last from fourteen to twenty years, there is little probability that they will be supplanted by iron sleepers at double the cost. But abroad the circumstances of cost and durability are different, and there the rolled iron or steel sleepers, which will outlive two or three sets of wooden ones, must claim advantages which cannot be overlooked. The difficulty will be in the fastenings, the mode of attaching the rails to the sleepers. The constant hammering of metal upon metal, resulting from the vibrations of every passing load, will quickly wear or loosen bolts, rivets, or wedges, and the fastenings which will prove the most efficient will be those that are the simplest and most readily adjusted.

Fastenings.—Figs. 320 to 335 illustrate some types of the principal fastenings used in connection with the chair road, and with flat-bottomed or flange rails.

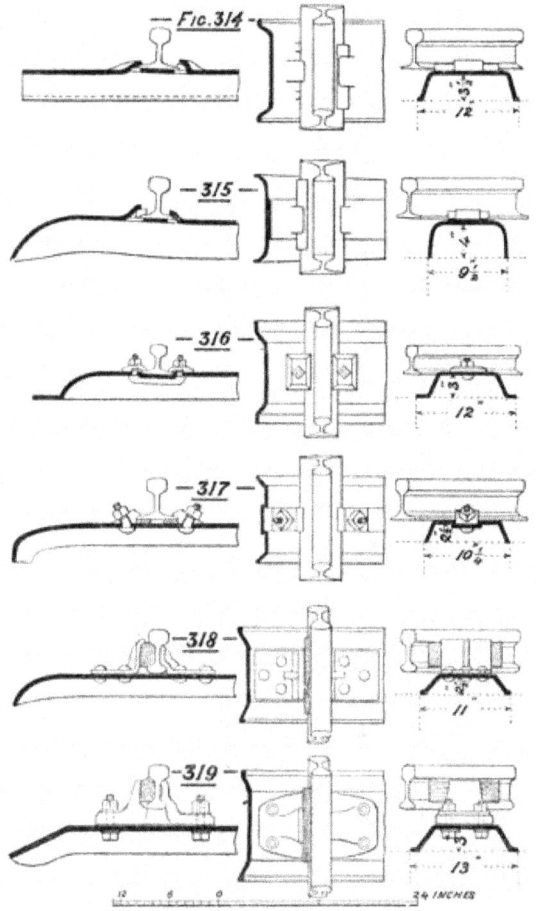

The fish-bolts, Figs. 320 and 321, are of a form which is in very general use both for steel bull-head rails and steel flange rails. By making the neck square or pear-shaped, to fit into corresponding hole in the fish-plate, the bolt is prevented from turning round when the wrench or spanner is applied to tighten the nut. A channel or groove is sometimes rolled on the outside of fish-plate to grip bolts made with square heads. Some engineers adopt two nuts, others prefer one nut of extra depth. Washers are used in some cases, but are not universal. With a deep rail it is preferable to place the nuts inside, so that the platelayer inspecting his length can see both rows of nuts as he walks along between the rails. With shallow rails the nuts must be placed outside and the cup-heads inside, to give ample clearance to the wheel-flanges.

Fish-bolts are subject to very severe work. Heavy rolling loads passing over the rail-joints—frequently at very high speeds—bring into play all the gripping power of the fish-bolts to maintain a firm support of the fish-plates to ends of rails, and the constant action of pressure and release produces a loosening or unscrewing motion in the bolts which is very difficult to counteract. Loose fish-bolts cause clattering joints and uneven road, and unless promptly remedied, the screw threads are soon destroyed and bolts rendered useless. Many devices have been invented to prevent or check this loosening of the bolts; one of the methods, and a very simple one, consists of a plain steel bolt with a steel lock-nut, made as shown in Fig. 322. As will be seen from the section, one-half of the nut is tapped of the same size as the bolt, and the remainder with deep-locking threads. The first half of the nut is readily screwed on to the bolt, but considerable force must be exerted to screw on the portion having the deep-locking threads; practically the second half of the nut has to cut a new or deeper thread for itself when screwing round the bolt.

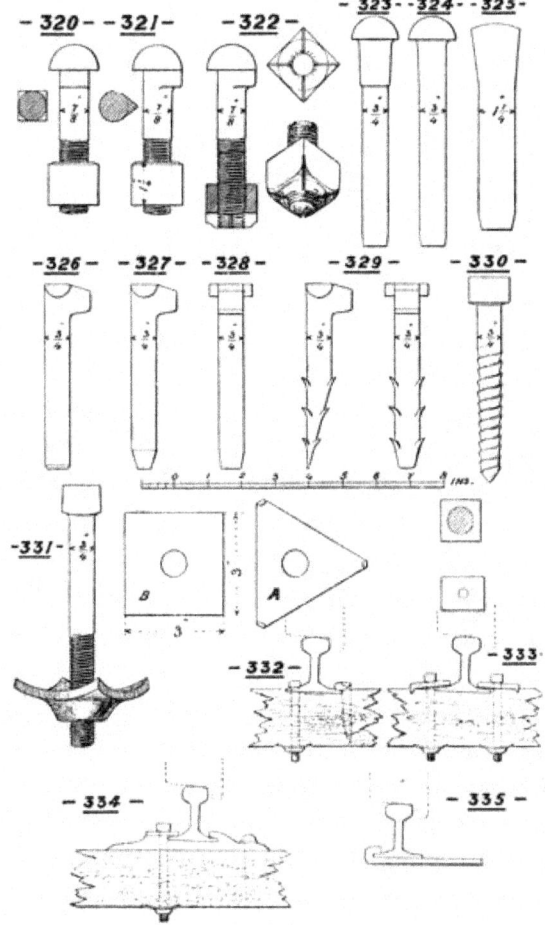

The slits or grooves at the angles of the nuts form four distinct cutting edges for shaping the deep threads. As the upper part of the lock-nut is divided by the grooves into four separate or detached segments, these segments will be forced slightly open or outwards during the action of cutting the deep thread on the bolt, and from their natural tendency to return to their original position they must exercise a strong gripping power on the bolt. This combined operation of cutting the deep threads and of forcing open the upper or detached segments, give an enormous holding and retaining power to the lock-nut, and enables it to withstand the train vibrations for a very long time without any perceptible slackening. In case of line repairs the nut can be readily unscrewed, and taken off the bolt.

Round iron spikes, as in Figs. 323 and 324, and round wooden trenails, as in Fig. 325, are both used for fastening cast-iron chairs to the sleepers. The spikes are made with a slightly taper neck, of size rather less than the hole in the chair, to avoid risk of breaking the casting when driving the spike down. Trenails are made out of well-seasoned hard wood, and are compressed by machinery. When driven into the sleeper, they expand by exposure to the atmosphere, and hold the chair very securely in position; but being only wood and of very small scantling, they are subject to early decay. The head, which is the only part in sight, may be perfectly sound, while the part between the chair-seat and top of sleeper may be quite rotten and useless. It would be very risky to depend upon trenails alone; one spike at least should be used to every chair. In some cases an extra large trenail is used with an augur-hole down the centre, through which either an iron spike is driven or a bolt is passed and screwed into a crab-nut on the under side of the sleeper. This arrangement will work well for a time, but there will be a great deal of play in the spike or bolt when the trenail becomes much decayed.

The spikes represented in Figs. 326, 327, and 328, are much used with flange rails. They are square in section, and finished with either blunt or sharp points, as shown. The top of spike is made with a doghead and side-lugs to facilitate the easing or withdrawal when necessary for renewals of sleepers, or alterations in line. By inserting the curved double claw end of a platelayers' crowbar, the spike can be raised without injuring the sleeper; but if it is required to be driven into the same sleeper again, a new hole must be bored, as the old hole will be too slack to be of any service. Augur-holes must be bored in the sleepers for the above spikes. For new roads, these holes can be bored by machinery when cutting the grooves for rail-seats; but when carrying out alterations or repairs, a large number of spike-holes must be bored by hand-augurs, an operation both slow and laborious. With the hand-boring there is the danger that the hole may not be made deep enough, owing to the workman's endeavour to avoid damaging the point of his augur by forcing it entirely through the sleeper, and bringing it in contact with a stone. Augur-holes bored wide to gauge will remain out of gauge, and although the spike may be driven down firm in its position, a space will be left for play between the rail-flange and spike.

Fig. 329 is a sketch of a dog-spike for flange rails which the writer has used for many years both abroad and at home, and which can be driven without any boring at all. The back of this spike is made perfectly straight, half of the front side is made parallel to the back, and the remainder is tapered down to a chisel point not exceeding 1/16 of an inch thick, the entering edge on the face being narrowed down to 3/8 of an inch in width. Three jags or spurs are cut on each side of the tapered portion, or twelve in all, and add greatly to the holding power. Not only can this spike be driven without any boring, but it possesses the additional advantage that in driving it down its taper or wedge-like shape causes it to drift hard up to the edge of the flange of rail, an element of great value in securing the exact gauge of line. With these spikes permanent-way laying can be carried on very rapidly, and they are especially valuable when making alterations, as augurs for spike-boring can be dispensed with altogether.

Wood screws with square heads similar to Fig. 330 are sometimes used for fastening flange rails to wooden sleepers. They are passed through holes punched or drilled in the flanges of the rails, and are intended to preserve the gauge as well as secure the rails to the sleepers. Experience has shown that these wood screws possess very limited holding power. The screwed portion of the bolt cuts but a very imperfect and weak holding thread in the soft wood of an ordinary sleeper, moisture insinuates itself into the bolt-hole, rusting the bolts and decaying the surrounding timber, and in a very short time the bolts become loose and incapable of holding the rail down firmly. As permanent-way fastenings wood screws are very inferior to crab bolts.

Crab bolts, as in Fig. 331, may be made either with square or hexagonal heads, and with three spur-nuts or four spur-nuts, as in **A** or **B**. The length of the bolts will depend upon the thickness of the sleeper or timber-work through which they have to be inserted. The bolt is pushed down through the hole bored in the sleeper, and the crab-nut put on from underneath. With a few turns of the bolt, the crab-nut is brought close up to the under side of the sleeper, the spur-points become embedded in the wood, and hold the nut firmly in position during subsequent tightening of the bolt. Crab bolts are extensively used with flange or flat-bottomed rails, and also in switch chairs and in crossings. A large number of flange rails are used with one hole through the flange at each end of rail, and a crab bolt passed through the hole and through the sleeper next to the joint, as shown in Fig. 332. This system checks the creeping of the rails by effectually securing or anchoring each rail to two of the sleepers. As there is always a tendency for these rails to crack through to the outside at the flange-holes, it is very desirable to have as few holes as possible. The two above described will be found sufficient for all practical purposes. To avoid punching or drilling more holes in the flanges of the rails, additional or intermediate crab bolts can be used by means of the fang clips shown on Fig. 333. The crab bolt is passed through the fang clips and through the sleeper close up to the flange of rail, and by screwing it round in the crab-nut under the sleeper the fang-clip is pressed down until the two spurs are driven into the timber, and the rail held securely in its place and to gauge. Intermediate crab-nuts and fang-clips should always be used in pairs, one on each side of the rail. Possessing more holding-down power than ordinary spikes, they are particularly valuable on sharp curves.

In some cases flange rails are laid in small cast-iron saddles, or chairs, as shown in Fig. 334, one end of the rail-seat having a recess to prevent the rail tilting upwards and outwards. An ordinary spike may be used for the inside end of chair, and a crab bolt with bent washer for the other. Unless the fastenings can be kept always tight, the above arrangement makes a very noisy, clattering road, as there are so many metal surfaces in contact, and so little to deaden the vibration. For narrow flange rails carrying heavy rolling load, chairs may be necessary to increase the bearing surface on the sleeper, but with rails having flanges five inches wide and upwards, it is better to let the flange rest direct on the wood of a properly grooved sleeper, and thus obtain a smoother and less noisy road.

On exceptionally sharp curves, wrought-iron or steel tie-bars, as in Fig. 335, are sometimes used to maintain the line to gauge. They may be made out of bars 3 inches wide by ½ an inch thick, turned over at the ends to grip the outside flanges. Being made to exact template, they have to be threaded on to the rails before spiking down, and are placed between the sleepers at distances from 7 to 10 feet apart.

Laying Permanent Way.—To preserve a good line and level to the permanent way, it is absolutely necessary that the road-bed should be kept thoroughly drained. If provision be not made for quickly carrying away the rain-water, and if it be allowed to collect under and around the sleepers, the action of the passing trains will work the finer particles of the packing into the consistency of soft mud, which will be gradually squeezed away, leaving the sleepers imperfectly supported and insecure. A loose sleeper involves a depression in the rails, and a corresponding lurch in the vehicles of the train, and a series of these depressions may produce such an oscillation in the train as to cause it to leave the rails.

The height or space from formation-level to rail-level is generally about 1 foot 9 inches for a flange railroad, and about 2 feet for a chair railroad.

Figs. 336 and 337 show cross-sections of both descriptions of road as laid down for a double line in cutting. The same arrangement applies to similar roads laid down in embankment, merely omitting the side-drains or water-tables. The bottom layer of ballast or road-bed should consist of good hard, quarried, or broken stones, each 6 inches deep, set on edge, firmly and closely hand-packed, forming a foundation through which the rain-water can be quickly carried away. On the top of this bottom pitching should be placed a 6-inch layer of broken stone ballast or strong clean gravel, of which none of the stones should be larger than will pass through a 2-inch ring. When the sleepers and rails have been laid on this second layer, and properly packed to line and level, the top ballasting, or boxing, of either broken stones or strong clean gravel, should be filled in to the form and extent specified. Where broken stones are used for the top ballasting none of them should be larger than will pass through a 1½-inch ring.

Broken stone ballast should only be made from the hardest and soundest description of rock or boulders, so that, however small the particles, they will remain sharp and clean.

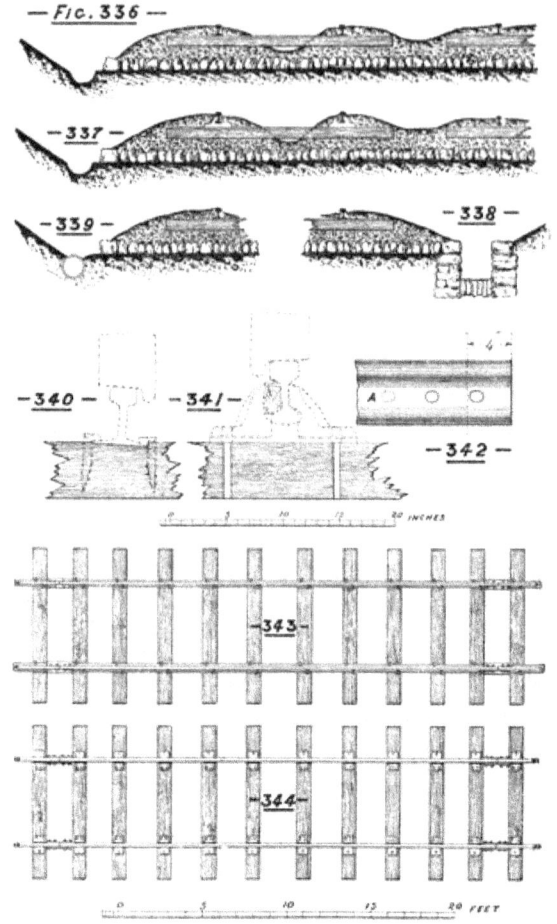

There are many kinds of rock which appear hard and compact when first excavated, but upon exposure to the weather undergo a complete change, developing into soft masses containing too much clay to allow the water to pass through readily. Where rock is scarce and gravel plentiful, the lower layer may be made of the heavier or coarser gravel, leaving the finer gravel for the upper layer, or boxing; but there is no doubt that the broken stone pitching makes the most efficient bottom layer. No gravel ballast should be used which is not free from clay or earthy sand.

Wherever there are particles of earthy matter, sufficient to furnish nourishment for vegetable growth, weeds will quickly spring up, and once established are most difficult, if not impossible, to eradicate.

The presence of weeds checks drainage, and gives an untidy appearance to the line, besides constantly occupying a large portion of the platelayers' time in their removal.

Clean cinders, free from dust or earth, are much used for upper ballast and boxing, and being lighter than gravel, are specially applicable for soft boggy ground. Burnt clay, broken into small pieces, has been largely adopted in districts where both rock and gravel were difficult to obtain. Chalk, furnace-slag broken small, crushed brick and sand, are frequently used as ballast. Sand is objectionable where there is high-speed traffic, as the finer particles rise in the form of dust and deposit themselves on the vehicles and machinery of the train.

The water-tables, or side drains in the cuttings, should be cut below the formation level, and to a depth or width sufficient to take away all rain-water, or water arising from springs. Where the material of the cutting is of a loose friable nature, it may be necessary to protect the sides of the water-tables with low dry stone walls, as in Fig. 338; or glazed earthenware pipes may be laid, as in Fig. 339, with open joints, or with grate openings at regular intervals. In some cases substantial side-walls and invert are requisite to carry away the flow of water.

Timber sleepers intended for the flange railroad should have the rail-seats grooved by machinery to ensure perfect accuracy in the position of the grooves, and in the angle or inclination of the rail-seats. Fig. 340 is a side view of part of a sleeper grooved to receive a flange rail. The presence of the grooves materially facilitate the laying of the rails to gauge, but must not be allowed to interfere with the constant use of the platelayer's gauge. In a similar manner the timber sleepers for the chair road frequently have the spike-holes bored to template by machinery, as indicated on Fig. 341. Steel or iron sleepers are delivered with the recesses for rails, and holes for bolts or fastenings formed complete by machinery.

The distances apart of the sleepers will be regulated in a great measure by the weight of the rails and the description of the traffic. Where light rails are intended to carry heavy engines the sleepers must be laid closer together than would be necessary for heavy rails. The joint being the weakest part of the rail, it is usual to put the sleepers closer together at that place, with a view to gain additional support, to assist the fish-plates in preserving as much as possible a firm unyielding surface at the rail-joint.

Fig. 343 shows an arrangement of sleepering largely adopted for steel flange rails 26 feet long, and weighing 79 lbs. per yard. The length of a rail is more a question of convenience of handling, facility of transhipment, and general use, than of actual manufacture. There is no difficulty in rolling rails up to 50 feet in length, or more; but very long rails are extremely ungainly things to move about, and are more exposed to receive permanent bends or kinks in unloading, besides requiring greater spaces at the joints to allow for contraction and expansion.

Fig. 344 is an example of sleepering for a chair railroad, for steel bull-head rails 26 feet long, and weighing 85 pounds per yard.

Line stakes and level pegs must be put in at suitable distances to guide the platelayers in laying the rails to the correct line and level, and on the curves the proper amount must be marked off for the super-elevation of the outer rail.

When the second layer of ballast has been spread for its full width and depth the sleepers can be distributed, and the rails or chairs spiked down to the correct gauge. Before putting on the fish-plates spaces must be left at the ends of the rails to allow for contraction and expansion, the amount depending upon the temperature at the time of laying down the rails. As the rails will expand, or increase in length, with the heat, it is necessary to allow more space for expansion for rails laid down in the cold, or winter months. On our home railways rails are very rarely laid down when the temperature is lower than 25° F., or higher than 125° F., and this range of 100° may be considered as covering all the variations likely to occur in ordinary practice. The greater portion of the permanent-way laying is carried on when the temperature is between 40° and 75°. The results of very carefully conducted experiments show that an increase of temperature of 1° F. will cause an iron or steel bar, or rail, to expand or lengthen to the extent of seven one-millionths of its length. Working this out for a range of 100° F. would give an increase in length of seven hundred one-millionths, which would be equal to an extension of 0·2184 of an inch in a 26-foot rail. For our home railways, therefore, a space of 5/16 of an inch will be found amply sufficient to meet the variations in length between the extremes of winter and summer, for a rail from 26 feet to 30 feet in length. Too much allowance for expansion is detrimental to the rails, because where the spaces are excessively large the wheels drop into the hollow and hammer or spread the ends of the rails.

The fish-bolts should not be completely tightened up until the permanent way is thoroughly set, and packed to its finished line and level.

On straight line the rail-joints should be laid square and opposite to each other. Permanent-way laying with broken joints is rarely adopted, except on curves or station-yards.

On curves the joints of the inner rails gain on the joints of the outer rails to the extent of—

$$\frac{\text{radius} + \text{gauge}}{\text{radius}} \times \text{length of rail}.$$

The amount of this gain, or lead, is adjusted by cutting off a portion of the end of the inner rails at certain intervals.

Assuming the fish-bolt holes to be spaced as shown on Fig. 342, then, when the inner rail is leading to the extent of 2 inches, a piece 4 inches long is cut off, as shown by dotted lines, leaving the original second fish-bolt hole to serve as first or end fish-bolt hole, and a new or second bolt-hole is drilled by hand at A. This method sets back the joint 2 inches from the square, and the lead is allowed to go on again until it becomes necessary to cut off another piece of 4 inches. Another mode is to have a proportion of the rails rolled 2 or 3 inches shorter for use on the curves.

On curves of a 1000 feet radius and upwards, the rails should be laid to the normal gauge, but on curves of lesser radius the gauge may be slightly increased, and as much as ¾ of an inch allowed on a curve of 500 feet radius.

The amount of cant, or super-elevation, to be given to the outer rail on curves must be regulated by the speed of the train and the gauge of the line. Many formulæ have been compiled to determine the necessary amount of super-elevation, but experience has shown that by some of them the calculated amounts were excessive. Possibly during past years too much cant has been given in many cases. The following simple formula approaches very closely to practical experience—

$$\frac{(\text{velocity in miles per hour})^2 \times \text{gauge in feet}}{\text{radius in feet} \times 1\cdot 25} = \text{the super-elevation of outer rail in inches.}$$

For high-speed trains uniformity of cant is of the utmost importance, more so even than the exact amount. Any irregularity in the super-elevation of the outer rail, sometimes high and sometimes low, will produce a dangerous swaying movement in the train, which, if not promptly checked, would lead to derailment.

More injury is done to curves by spreading, arising from rigid wheel-bases of engines and tenders, than from any want of counteraction to centrifugal force.

When a long length of permanent way has been linked in, rails spiked to gauge, and fish-plates bolted together, the platelayers can proceed to the final adjustment to line and level in accordance with the stakes and pegs provided for their guidance. The setting to exact line is effected by means of long pointed round iron crowbars, which are struck forcibly into the ballast alongside the rails, and serve as powerful hand-levers to pull or push the rails to the right or left as directed by the foreman standing some distance back at one of the line-stakes. The men with the crowbars pass from rail-length to rail-length, until a long stretch of road has been pulled into correct line.

The adjustment to rail-level is done by first packing up the sleepers to the correct height at the various level-pegs, and then packing up the intermediate sleepers so that the surface of the top of the rails forms one uniform even line from level-peg to level-peg. On new lines it is usual to pack a little high in the first instance to allow for the subsidence or compression which invariably takes place on the passage of heavy trains over fresh ballast.

The form or contour line of the top ballast will vary according to circumstances. In station-yards it is usual to fill in the ballast almost up to the level of the top of the rails for the convenience and safety of the men who are constantly moving about marshalling the carriages and waggons. Out on the open line between stations, the ballast on some railways is filled in up to rail-level, while on others it is only filled in up to the tops of the sleepers, leaving the rails and chairs quite clear of the ballast. On others, again, the ballast is filled well up to the rails and channelled in the centre, as shown on the sketches Figs. 336 and 337. Channelling the centre of the road reduces the quantity of ballast per mile, ensures good drainage, and also stability by not permitting any central support to the sleepers. By covering up the lower table and sides of rails the noise is reduced to a minimum, vibration is absorbed, and a more silent road is the result. The contact with the ballast also preserves the rail from the extremes of temperature. Where the ballasting is not channelled there is some risk of the sleepers breaking in the middle. The constant packing of the sleepers just under the rails has a tendency to drift some of the ballast inwards towards the middle of the sleeper, forming a hard compact mass, and this mass, acting as fulcrum, throws considerable strain on the middle of the sleeper when the trains pass over and depress the ends. Where the ballast is filled in level with the rails on top of sleepers it should be loosened occasionally in the middle to prevent it becoming too hard.

Connections with the rails of the main line will have to be made in various forms to suit the circumstances of the joining lines or sidings.

Fig. 345 shows a simple double-line junction.

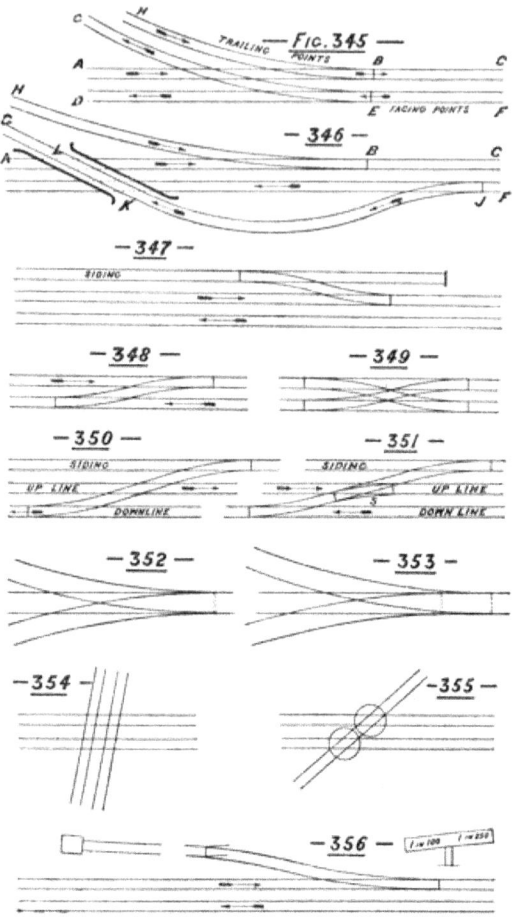

Fig. 346 shows an example of what is termed a *flying junction*, or a junction of two double lines arranged in such a manner as to cause the least interruption to a constant train traffic passing UP and DOWN over both lines. Upon referring to Fig. 345 it will be seen that a train from **F**, turning off at the points **E** and proceeding to **G**, must block, or close for traffic the section **ABC** during its passage over that line towards **G**. With a crowded train-service the blocking of both UP and DOWN main lines for the working of one train would cause much interruption, and to obviate such delay the *flying junction* is substituted. Fig. 346 shows how a train from **F** is turned off at the points **J** and proceeds on to **K**, where by means of a bridge it passes either over or under both main lines, and continues on to **G** without in any way interfering with the train service on **ABC**.

Fig. 347 is an ordinary plain siding or *turn-out*, including the necessary throw-off or trap-points and short dead end.

Fig. 348 is an ordinary *cross-over road* from DOWN main line to UP main line, and *vice versâ*.

Fig. 349 is a double cross-over road, generally termed a *scissors* cross-over.

Fig. 350 is a simple through cross-over road from DOWN main line to siding alongside UP main line.

Fig. 351 is a similar arrangement of through cross-over road with the addition of a pair of slip points at **S** to make a connection with the UP main line, thus combining the facilities of the ordinary cross-over and through cross-over road.

Fig. 352 shows a set of three throw-switches with all the sliding tongues placed side by side; and Fig. 353 shows another arrangement of three throws with the sliding-rails of the second set of switches placed just behind the heel of the first set of switches. The latter method works very well where there is sufficient length for the purpose.

Fig. 354 shows a square crossing, where one line of railway crosses another line of railway on the same level.

Fig. 355 shows a connection with a siding by means of an ordinary carriage or waggon turn-table.

Fig. 356 shows a set of "runaway" points which are sometimes placed in the main line at the top of an incline close to a station, the object being to intercept or throw off any portion of a train which may have become detached, and which would, if unchecked, run away back down the incline. By means of a weighted lever or spring the points are set to the normal position of *open* to the siding, and as they are "trailing" points for the running road they are readily closed by a passing train. One or other of the above forms of connections, or a combination of them, will meet all the requirements which usually occur in railway work.

Fig. 357 is an enlarged sketch of an ordinary cross-over road, and Fig. 358 of a double or *scissors* cross-over.

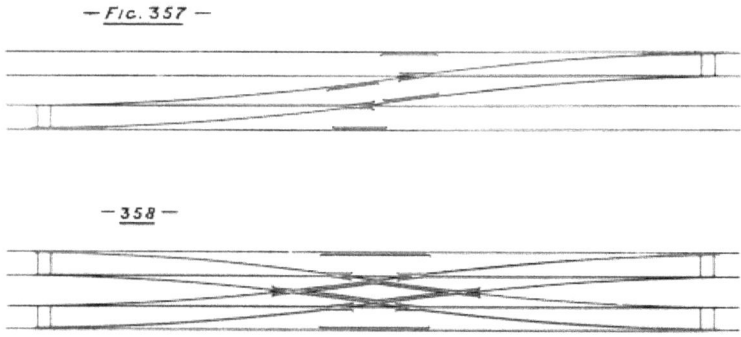

Fig. 359 shows a *single-slip* point connection, and Fig. 360 a *double-slip* point connection. In places where slip connections can be introduced they add greatly to the facilities for train movements without curtailing the available standing-room for vehicles on the lines and sidings. They are simple in construction, do not require crossings, and in many cases save a complete cross-over road. At the same time slip connections can only be laid down where the angle of the intersecting lines is sufficiently flat to admit of a connecting curve of workable radius.

Fig. 361 is an enlarged sketch of a set of ordinary 15-foot switches or points. By placing them about the middle of the stock rails the joints of the latter are kept well beyond the sliding rails, and the road is held firmly together. It is necessary to place the sleepers closer together at the switches to allow for the reduction in section of the sliding rails, which results from planing them down to the requisite shape. By substituting two long timbers for the ordinary sleepers at the points of the switch rails, as shown on the sketch, a more efficient support is obtained for the switch-box or crank in the case of rod-worked switches, and the working distance from the rails is accurately maintained, irrespective of any packing or pulling of the road. In the sketch a steel bull-head rail is shown on one side, and a steel flange rail on the other, each bolted to an ordinary cast-iron switch chair. Switch chairs are sometimes made of plates of wrought-iron or steel, forged to the correct shape, and riveted together. They are, however, much more costly than cast-iron chairs, and deteriorate more quickly from corrosion.

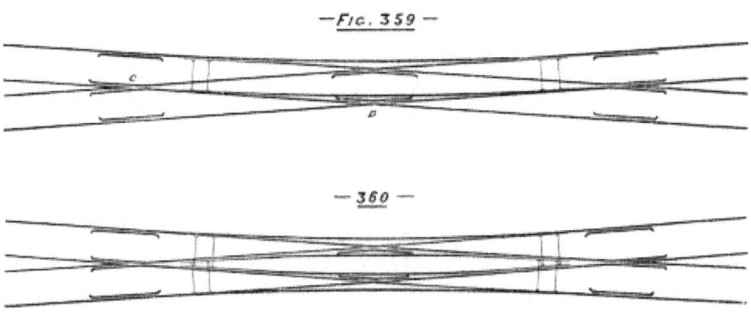

Fig. 362 is an enlarged sketch of an ordinary crossing similar to the one indicated at **C** (Fig. 359), and composed of a cast-steel reversible block. The ends and lugs, **L, L**, are formed to suit the connecting rails and fish-plates, as shown in the cross-sections. The casting is secured to the crossing timbers by bolts passing through the side lugs, **S**, a cast-iron packing-washer, **W**, being placed between the lug and the timber to ensure a solid seat and avoid rocking. A very important point in the construction of these block crossings is to have the groove or flange-path sufficiently deep to prevent the striking or touching of the flange of a much-worn tyre. A well-made, carefully annealed steel-block reversible crossing is very smooth in the road, and has a long life. It is all in one solid piece; there are no parts to work loose or spread; the wear of the running surface is very uniform, and when the one side is much worn down, there is the other ready for service. The writer has had many of these steel-block reversible crossings in use under heavy and fast traffic for six and eight years without turning.

Fig. 363 shows an ordinary crossing made of steel bull-head rails secured in strong cast-iron chairs; and Fig. 364 is a similar crossing made of steel flange rails. In some cases the two rails forming the **V** are welded together at the point **B**, and in others they are riveted or bolted together. Fig. 365 shows a diamond or through crossing similar to the one indicated at **D**, Fig. 359, made of steel bull-head rails and chairs.

Crossings are constructed in a variety of forms, whether on the principle of the cast-steel block, or made out of ordinary steel rails; and the above sketches merely illustrate some well-recognized types which experience has proved to be efficient and durable in the road. The angles of the crossings will depend upon the divergence of the intersecting lines to be connected; ordinary crossings, to the angle of 1 in 10, work in for very general use in station-yards, but many are required of angles varying from 1 in 6 to 1 in 14, and in some cases 1 in 16.

As a rule, engineers endeavour as far as possible to avoid using ordinary crossings flatter than 1 in 12, or diamond crossings flatter than 1 in 9, because the gap between the running rails becomes very considerable beyond those angles. At the same time, there are many cases of ordinary crossings of 1 in 16, and diamond crossings of 1 in 12 and 1 in 13 laid down in exceptional places, and which have carried heavy and fast traffic for many years. All crossings should be well protected with wing rails and guard rails, as shown on the sketches.

Fig. 366 illustrates a method of bringing the UP and down lines of a double line of railway close to each other, and passing them over a single-line opening bridge, or a bridge where the works for the second line have not been completed. This arrangement avoids the necessity of any switches, and prevents any accidents which would arise from a misplaced switch. Each set of trains is effectually kept to its own line of rails. With proper signalling or pilot working, the double-line traffic can be worked over the single-line bridge without difficulty. The writer has adopted the above arrangement in many cases when renewing double-line bridges or viaducts where the width for traffic working has been restricted to half of the bridge.

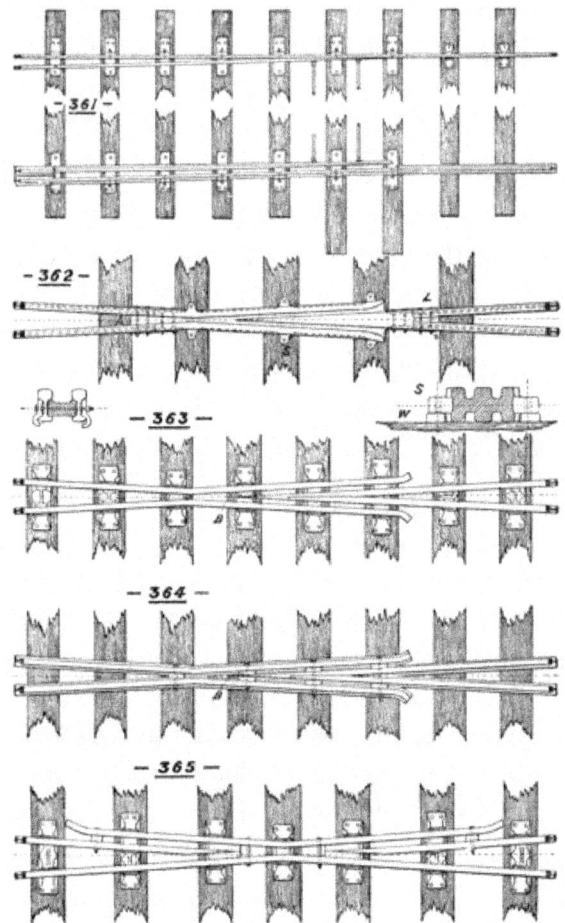

In some instances the same system has been extended to the carrying of four lines of rails over a double-line bridge, as shown on Fig. 367.

The principal tool used by platelayers for lifting the permanent way is a long iron-shod wooden lever, as shown in Fig. 368. The point of the lower end is pushed under the sleeper, and the curved shoulder placed on a large stone or piece of wood as a support, and then by pulling down the upper end of the lever the road can be lifted to the height required. Screw lifting-jacks of various kinds are also used for the same purpose, the foot or base of the jack resting on the ballast, while the claws grasp the under side of the rail, and raise it by means of the screw. With appliances which lift by the rails, the sleepers have to be raised by the holding power of the spikes or bolts, an operation which is apt to throw undue strain on spikes. Where possible it is preferable to lift from the under side of the sleepers.

Beaters similar to the one shown on Fig. 369 are used for packing the ballast. One end of the beater is pointed like a pick, and serves to loosen the ballast or broken stone, and the other end is made somewhat in the hammer-head form to pack or beat the ballast under the sleeper. With skilled men the beater is a most useful tool, speedy and effective in its action. Held in both hands, it is raised slightly, and then brought down sharply, the hammer-head striking the gravel or broken stone placed alongside for packing under the sleeper. A series of smart blows can be given with rapidity and without requiring any great muscular effort. In some foreign countries there is difficulty in initiating the natives to work with the ordinary beater, on account of the stooping position necessary for its use. To meet this difficulty the writer has in many cases substituted a packing or tamping bar, as shown in Fig. 370. This bar, about 5 feet long, is made of light round wrought-iron or steel, with a ring-shaped handle at one end, and an ordinary beater head at the other. The workman using this bar stands upright, guides the bar, held loosely, with his left hand, and with his right gives a continuance of smart blows. This tool works well in the hands of light active natives, who can thus give a number of rapid strokes without much exertion.

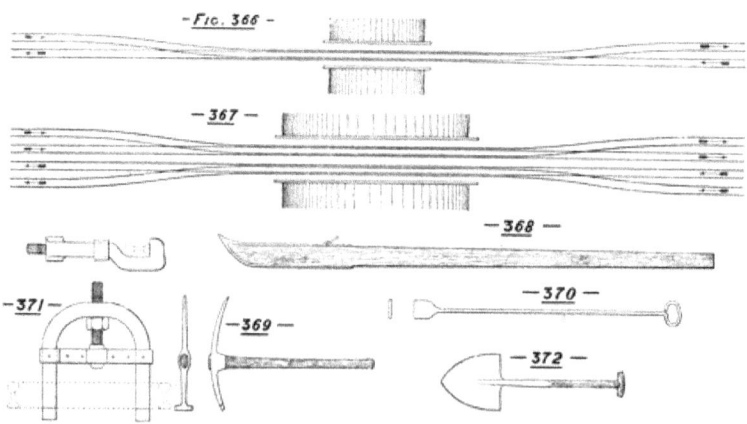

The simple rail-bender, or *Jim Crow*, of the form shown in Fig. 371, is much used by platelayers for giving a slight bend or set to rails which have to be laid down on sharp curves on main line or cross-over roads. The rail is laid across the two arms, and the screw turned round and downwards by means of an iron bar lever used as a spanner or wrench to the nut shown on the sketch. The same tool is also serviceable for straightening rails which have become crooked or kinked. Large and more comprehensive machines are used for bending rails in large quantities or setting them to exact curvature, but, being heavy and cumbersome, they are rarely taken away from the store-yards.

Strong steel shovels of the form shown in Fig. 372 are the most suitable for platelayers' general use when working with gravel, sand, or broken stones.

For driving iron spikes and wooden keys in cast-iron chairs a long-handled hammer is the most convenient for work, and its long swinging action produces considerable force without much actual labour.

Road-gauges, nut-wrenches, short straight-edges, spirit-levels, ratchet-drills, augurs, and cold setts of well-tempered steel for cutting rails, are all required by the men engaged in laying permanent way.

The following summaries give the estimated cost of materials alone for one mile of steel bull-head rail and steel flange rail permanent way of different weights. The 90-lb. steel bull-head rail is at present the heaviest of that section laid down to any extent on our home railways, and the chairs and fastenings are made heavy to correspond to the rail and the traffic for which it is intended. As the rails in the summaries become lighter, the weights of the chairs and fastenings are decreased. As yet there are not many samples of the 100-lb. steel flange rail; but in those places where it has been laid down it has been supported with a liberal supply of sleepers, to obtain increased bearing surface. With a 5½-inch flange, and a rectangular sleeper 10 inches wide, the bearing surface on the wood is only about 55 square inches, as compared with about 100 square inches, the bearing surface of a large cast-iron chair for a heavy bull-head rail. As previously explained, a small bearing surface on a sleeper tends to the cutting down into the wood, and rendering the sleeper unsafe and useless even before it has become unserviceable from decay: hence the reason for ample bearing surface on the sleeper. The last two summaries refer to 3-foot narrow-gauge lines. In more than one instance the 45-lb. rails first laid down have been found much too light for the engines required to work the traffic, and when making extensions of the system 65-lb. rails have been adopted. Indeed, when taking into consideration the weight of most of the narrow-gauge engines, generally from 24 to 28 tons in working order, and their short wheel-base, it would appear that a 65-lb. rail is the minimum which should be used both for stability and economy in maintenance.

The summaries are prepared from examples in actual use, and represent the number and weight of sleepers, chairs, and fastenings in each instance. Even with the same weight of rail, the practice differs on various lines as to the weights of the chairs and fastenings; and the selections have been made to show a fair average. On some railways the chairs are secured partly by tree-nails and partly by spikes, or crab bolts; on others only spikes are used. The prices put down are the estimated values of the materials delivered into the Permanent Way Stores of our own home railways, and are exclusive of all costs of freight, carriage, or distribution to the site of laying down. The prices are only comparative, and fluctuate up or down according to the current value of the raw materials from which the various items are manufactured. Lighter rails and smaller fastenings cost more per ton than those of a heavier type, as they involve more labour and workmanship.

STEEL BULL-HEAD RAILS (90 LBS. PER YARD).
Estimated Cost of Materials for One Mile of Single Line.

	Weight per mile of single line.			
	tons.	cwt.	qrs.	lbs.
Steel bull-head rails, 90 lbs. per yard (30-ft. lengths)	141	8	2	0
Steel fish-plates (deep), 41 lbs. per pair	6	10	0	0
Fish-bolts and nuts	1	4	0	0
2112 creosoted sleepers, 9 ft. × 10 in. × 5 in.	—			
4224 cast-iron chairs, each 50 lbs.	94	5	3	0
8448 iron cup-headed spikes	3	15	2	0
8448 tree-nails, at per 1000	—			
4224 oak keys, at per 1000	—			

	Price.			Amount.		
	£	s.	d.	£	s.	d.
Steel bull-head rails, 90 lbs. per yard (30-ft. lengths)	5	0	0	707	2	6
Steel fish-plates (deep), 41 lbs. per pair	6	15	0	43	17	6
Fish-bolts and nuts	12	15	0	15	6	0
2112 creosoted sleepers, 9 ft. × 10 in. × 5 in.	0	3	10	404	16	0
4224 cast-iron chairs, each 50 lbs.	3	10	0	330	0	1
8448 iron cup-headed spikes	10	0	0	37	15	0
8448 tree-nails, at per 1000	3	10	0	29	11	4
4224 oak keys, at per 1000	5	0	0	21	2	5
			£	1589	10	10

STEEL BULL-HEAD RAILS (85 LBS. PER YARD).
Estimated Cost of Materials for One Mile of Single Line.

	Weight per mile of single line.			
	tons.	cwt.	qrs.	lbs.
Steel bull-head rails, 85 lbs. per yard (26-ft. length)	134	0	0	0
Steel fish-plates (deep), 38 lbs. per pair	6	17	3	0
Fish-bolts and nuts	1	4	3	0
2030 creosoted sleepers, 9 ft. × 10 in. × 5 in.	—			
4060 cast-iron chairs, each 45 lbs.	81	11	1	0
8120 iron cup-headed spikes	3	12	2	0
4060 tree-nails, at per 1000	—			
4060 oak keys, at per 1000	—			

	Price.			Amount.		
	£	s.	d.	£	s.	d.
Steel bull-head rails, 85 lbs. per yard (26-ft. length)	5	0	0	670	0	0
Steel fish-plates (deep), 38 lbs. per pair	6	15	0	46	9	10
Fish-bolts and nuts	12	15	0	15	15	7
2030 creosoted sleepers, 9 ft. × 10 in. × 5 in.	0	3	10	389	1	8
4060 cast-iron chairs, each 45 lbs.	3	10	0	285	9	5
8120 iron cup-headed spikes	10	0	0	36	5	0
4060 tree-nails, at per 1000	3	10	0	14	4	2
4060 oak keys, at per 1000	5	0	0	0	6	0
			£	1477	11	8

STEEL BULL-HEAD RAILS (80 LBS. PER YARD).
Estimated Cost of Materials for One Mile of Single Line.

	Weight per mile of single line.			
	tons.	cwt.	qrs.	lbs.
Steel bull-head rails, 80 lbs. per yard (26-ft. lengths)	125	14	0	0
Steel fish-plates (deep), 37 lbs. per pair	6	14	1	0
Fish-bolts and nuts	1	4	3	0
2030 creosoted sleepers, 9 ft. × 10 in. × 5 in.	—			
4060 cast-iron chairs, each 40 lbs.	72	10	0	0
8120 iron cup-headed spikes	3	12	2	0
4060 tree-nails, at per 1000	—			
4060 oak keys, at per 1000	—			

	Price.			Amount.		
	£	s.	d.	£	s.	d.
Steel bull-head rails, 80 lbs. per yard (26-ft. lengths)	5	0	0	628	10	0
Steel fish-plates (deep), 37 lbs. per pair	6	15	0	45	6	2
Fish-bolts and nuts	12	15	0	15	15	7
2030 creosoted sleepers, 9 ft. × 10 in. × 5 in.	0	3	10	389	1	8
4060 cast-iron chairs, each 40 lbs.	3	10	0	253	15	0
8120 iron cup-headed spikes	10	0	0	36	5	0
4060 tree-nails, at per 1000	3	10	0	14	4	2
4060 oak keys, at per 1000	5	0	0	20	6	0
				£ 1403	3	7

STEEL BULL-HEAD RAILS (75 LBS. PER YARD).
Estimated Cost of Materials for One Mile of Single Line.

	Weight per mile of single line.			
	tons.	cwt.	qrs.	lbs.
Steel bull-head rails, 75 lbs. per yard (26-ft. lengths)	117	0	0	0
Steel fish-plates (deep), 35 lbs. per pair	6	7	0	0
Fish-bolts and nuts	1	4	3	0
2030 creosoted sleepers, 9 ft. × 10 in. × 5 in.	—			
4060 cast-iron chairs, each 37 lbs.	67	1	1	0
12,180 iron cup-headed spikes	5	8	3	0
4060 oak keys, at per 1000	—			

	Price.			Amount.		
	£	s.	d.	£	s.	d.
Steel bull-head rails, 75 lbs. per yard (26-ft. lengths)	5	0	0	585	0	0
Steel fish-plates (deep), 35 lbs. per pair	6	15	0	42	17	3
Fish-bolts and nuts	12	15	0	15	15	7
2030 creosoted sleepers, 9 ft. × 10 in. × 5 in.	0	3	10	389	1	8
4060 cast-iron chairs, each 37 lbs.	3	10	0	234	14	5
12,180 iron cup-headed spikes	10	0	0	54	7	6
4060 oak keys, at per 1000	5	0	0	20	6	0
			£	1342	2	5

Steel Bull-head Rails (70 lbs. per Yard).
Estimated Cost of Materials for One Mile of Single Line.

	Weight per mile of single line.			
	tons.	cwt.	qrs.	lbs.
Steel bull-head rails, 70 lbs. per yard (26-ft. lengths)	110	0	0	0
Steel fish-plates (deep), 32 lbs. per pair	5	16	0	0
Fish-bolts and nuts	1	2	0	0
2030 creosoted sleepers, 9 ft. × 10 in. × 5 in.	—			
4060 cast-iron chairs, each 34 lbs.	61	12	2	0
8120 iron cup-headed spikes	3	3	2	0
4060 oak keys, at per 1000	—			

	Price.			Amount.		
	£	s.	d.	£	s.	d.
Steel bull-head rails, 70 lbs. per yard (26-ft. lengths)	5	0	0	550	0	0
Steel fish-plates (deep), 32 lbs. per pair	6	15	0	39	3	0
Fish-bolts and nuts	12	15	0	14	0	6
2030 creosoted sleepers, 9 ft. × 10 in. × 5 in.	0	3	10	389	1	8
4060 cast-iron chairs, each 34 lbs.	3	10	0	215	13	9
8120 iron cup-headed spikes	10	0	0	31	15	0
4060 oak keys, at per 1000	4	10	0	18	5	5
			£	1257	19	4

STEEL BULL-HEAD RAILS (65 LBS. PER YARD).
Estimated Cost of Materials for One Mile of Single Line.

	Weight per mile of single line.			
	tons.	cwt.	qrs.	lbs.
Steel bull-head rails, 65 lbs. per yard (26-ft. lengths)	102	3	0	0
Steel fish-plates (deep), 28 lbs. per pair	5	1	2	0
Fish-bolts and nuts	1	1	0	0
2030 creosoted sleepers, 9 ft. × 9 in. × 4½ in.	—			
4060 cast-iron chairs, each 28 lbs.	50	15	0	0
8120 iron cup-headed spikes	2	19	0	0
4060 oak keys, at per 1000	—			

	Price.			Amount.		
	£	s.	d.	£	s.	d.
Steel bull-head rails, 65 lbs. per yard (26-ft. lengths)	5	5	0	536	5	9
Steel fish-plates (deep), 28 lbs. per pair	7	0	0	35	10	6
Fish-bolts and nuts	13	0	0	13	13	0
2030 creosoted sleepers, 9 ft. × 9 in. × 4½ in.	0	3	0	304	10	0
4060 cast-iron chairs, each 28 lbs.	4	0	0	203	0	0
8120 iron cup-headed spikes	10	10	0	30	19	6
4060 oak keys, at per 1000	4	0	0	16	4	9
			£	1140	3	6

STEEL FLANGE RAILS (100 LBS. PER YARD).
Estimated Cost of Materials for One Mile of Single Line.

	Weight per mile of single line.			
	tons.	cwt.	qrs.	lbs.
Steel flange rails, 100 lbs. per yard (30-ft. lengths)	157	3	0	0
Steel fish-plates (deep), 42 lbs. per pair	6	12	0	0
Fish-bolts and nuts	1	5	0	0
2464 creosoted sleepers, 9 ft. × 10 in. × 5 in.	—			
8448 dog-head spikes	3	6	0	0
704 fang clips	0	8	3	0
1408 crab bolts	1	6	3	0

	Price.			Amount.		
	£	s.	d.	£	s.	d.
Steel flange rails, 100 lbs. per yard (30-ft. lengths)	5	0	0	785	15	0
Steel fish-plates (deep), 42 lbs. per pair	6	10	0	42	18	0
Fish-bolts and nuts	12	15	0	15	18	9
2464 creosoted sleepers, 9 ft. × 10 in. × 5 in.	0	3	10	472	5	4
8448 dog-head spikes	12	10	0	41	5	0
704 fang clips	13	10	0	5	18	2
1408 crab bolts	12	10	0	16	14	5
			£	1380	14	8

STEEL FLANGE RAILS (79 LBS. PER YARD).
Estimated Cost of Materials for One Mile of Single Line.

	Weight per mile of single line.			
	tons.	cwt.	qrs.	lbs.
Steel flange rails, 79 lbs. per yard (26-ft. lengths)	125	0	0	0
Steel fish-plates (deep), 37 lbs. per pair	6	14	1	0
Fish-bolts and nuts	1	4	0	0
2030 creosoted sleepers, 9 ft. × 10 in. × 5 in.	—			
6496 dog-head spikes	2	10	3	0
812 fang clips	0	10	0	0
1624 crab bolts	1	10	3	0

	Price.			Amount.		
	£	s.	d.	£	s.	d.
Steel flange rails, 79 lbs. per yard (26-ft. lengths)	5	0	0	625	0	0
Steel fish-plates (deep), 37 lbs. per pair	6	10	0	43	12	8
Fish-bolts and nuts	12	15	0	15	6	0
2030 creosoted sleepers, 9 ft. × 10 in. × 5 in.	0	3	10	389	1	8
6496 dog-head spikes	12	10	0	31	14	5
812 fang clips	13	10	0	6	15	0
1624 crab bolts	12	10	0	19	4	5
			£	1130	14	2

STEEL FLANGE RAILS (74 LBS. PER YARD).
Estimated Cost of Materials for One Mile of Single Line.

	Weight per mile of single line.			
	tons.	cwt.	qrs.	lbs.
Steel flange rails, 74 lbs. per yard (30-ft. lengths)	116	5	3	0
Steel fish-plates (deep), 30½ lbs. per pair	4	15	3	0
Fish-bolts and nuts	1	1	0	0
1936 creosoted sleepers, ft. × 10 in. × 5 in.	—			
6336 dog-head spikes	2	9	2	0
704 fang clips	0	8	3	0
1408 crab bolts	1	6	3	0

	Price.			Amount.		
	£	s.	d.	£	s.	d.
Steel flange rails, 74 lbs. per yard (30-ft. lengths)	5	0	0	581	8	9
Steel fish-plates (deep), 30½ lbs. per pair	6	10	0	31	2	5
Fish-bolts and nuts	12	15	0	13	7	9
1936 creosoted sleepers, ft. × 10 in. × 5 in.	0	3	10	371	1	4
6336 dog-head spikes	12	10	0	30	18	9
704 fang clips	13	10	0	5	18	2
1408 crab bolts	12	10	0	16	14	5
			£	1050	11	7

STEEL FLANGE RAILS (65 LBS. PER YARD).
Estimated Cost of Materials for One Mile of Single Line.

	Weight per mile of single line.			
	tons.	cwt.	qrs.	lbs.
Steel flange rails, 65 lbs. per yard (30-ft. lengths)	102	3	0	0
Steel fish-plates (deep), 27 lbs. per pair	4	4	3	0
Fish-bolts and nuts	1	0	0	0
1936 creosoted sleepers, 9 ft. × 10 in. × 5 in.	—			
6336 dog-head spikes	2	9	2	0
704 fang clips	0	8	0	0
1408 crab bolts	1	6	3	0

	Price.			Amount.		
	£	s.	d.	£	s.	d.
Steel flange rails, 65 lbs. per yard (30-ft. lengths)	5	10	0	561	16	6
Steel fish-plates (deep), 27 lbs. per pair	7	5	0	30	14	5
Fish-bolts and nuts	13	0	0	13	0	0
1936 creosoted sleepers, 9 ft. × 10 in. × 5 in.	0	3	10	371	1	4
6336 dog-head spikes	12	10	0	30	18	9
704 fang clips	13	10	0	5	8	0
1408 crab bolts	12	10	0	16	14	5
			£	1029	13	5

Steel Flange Rails (60 lbs. per Yard).
Estimated Cost of Materials for One Mile of Single Line.

	Weight per mile of single line.			
	tons.	cwt.	qrs.	lbs.
Steel flange rails, 60 lbs. per yard (30-ft. lengths)	94	5	3	0
Steel fish-plates (deep), 25 lbs. per pair	3	18	2	0
Fish-bolts and nuts	1	0	0	0
2112 creosoted sleepers, 9 ft. × 10 in. × 5 in.	—			
7040 dog-head spikes	2	15	0	0
704 fang clips	0	8	0	0
1408 crab bolts	1	6	3	0

	Price.			Amount.		
	£	s.	d.	£	s.	d.
Steel flange rails, 60 lbs. per yard (30-ft. lengths)	5	10	0	518	11	7
Steel fish-plates (deep), 25 lbs. per pair	7	5	0	28	9	2
Fish-bolts and nuts	13	0	0	13	0	0
2112 creosoted sleepers, 9 ft. × 10 in. × 5 in.	0	3	10	404	16	0
7040 dog-head spikes	12	10	0	34	7	6
704 fang clips	13	10	0	5	8	0
1408 crab bolts	12	10	0	16	14	5
			£	1021	6	8

STEEL FLANGE RAILS (50 LBS. PER YARD).
Estimated Cost of Materials for One Mile of Single Line.

	Weight per mile of single line.			
	tons.	cwt.	qrs.	lbs.
Steel flange rails, 50 lbs. per yard (30-ft. lengths)	78	11	2	0
Steel fish-plates (deep), 22 lbs. per pair	3	9	1	0
Fish-bolts and nuts	0	18	0	0
2112 creosoted sleepers, 9 ft. × 9 in. × 4½ in.	—			
7040 dog-head spikes	2	7	1	0
704 fang clips	0	6	1	0
1408 crab bolts	1	2	0	0

	Price.			Amount.		
	£	s.	d.	£	s.	d.
Steel flange rails, 50 lbs. per yard (30-ft. lengths)	5	15	0	451	16	1
Steel fish-plates (deep), 22 lbs. per pair	7	10	0	25	19	5
Fish-bolts and nuts	13	10	0	12	3	0
2112 creosoted sleepers, 9 ft. × 9 in. × 4½ in.	0	3	0	316	16	0
7040 dog-head spikes	13	0	0	30	14	3
704 fang clips	14	0	0	4	7	6
1408 crab bolts	13	0	0	14	6	0
			£	856	2	3

STEEL FLANGE RAILS (65 LBS. PER YARD).
Estimated Cost of Materials for One Mile of Single Line (3-ft. gauge).

	Weight per mile of single line.			
	tons.	cwt.	qrs.	lbs.
Steel flange rails, 65 lbs. per yard (30-ft. lengths)	102	3	0	0
Steel fish-plates (deep), 27 lbs. per pair	4	4	3	0
Fish-bolts and nuts	1	0	0	0
2288 creosoted sleepers, 6 ft. × 9 in. × 4½ in.	—			
7744 dog-head spikes	2	17	0	0
704 fang clips	0	7	2	0
1408 crab bolts	2	2	0	0

	Price.			Amount.		
	£	s.	d.	£	s.	d.
Steel flange rails, 65 lbs. per yard (30-ft. lengths)	5	10	0	561	16	6
Steel fish-plates (deep), 27 lbs. per pair	7	5	0	30	14	5
Fish-bolts and nuts	13	0	0	13	0	0
2288 creosoted sleepers, 6 ft. × 9 in. × 4½ in.	0	2	3	257	8	0
7744 dog-head spikes	12	10	0	35	12	6
704 fang clips	13	10	0	5	1	3
1408 crab bolts	12	10	0	26	5	0
			£	929	17	8

STEEL FLANGE RAILS (45 LBS. PER YARD).
Estimated Cost of Materials for One Mile of Single Line (3-ft. gauge).

	Weight per mile of single line.			
	tons.	cwt.	qrs.	lbs.
Steel flange rails, 45 lbs. per yard (26-ft. lengths)	70	14	1	0
Steel fish-plates (deep), 16 lbs. per pair	2	18	0	0
Fish-bolts and nuts	0	18	0	0
2233 creosoted sleepers, 6 ft. × 8 in. × 4 in.		—		
7308 dog-head spikes	2	14	0	0
812 fang clips	0	5	0	0
1624 crab bolts	0	18	0	0

	Price.			Amount.		
	£	s.	d.	£	s.	d.
Steel flange rails, 45 lbs. per yard (26-ft. lengths)	5	15	0	406	12	0
Steel fish-plates (deep), 16 lbs. per pair	7	10	0	21	15	0
Fish-bolts and nuts	13	10	0	12	3	0
2233 creosoted sleepers, 6 ft. × 8 in. × 4 in.	0	1	10	204	13	10
7308 dog-head spikes	13	0	0	35	2	0
812 fang clips	14	0	0	3	10	0
1624 crab bolts	13	0	0	11	14	0
			£	695	9	10

CHAPTER IV.

Stations: Station Buildings, Roofs, Lines, and Sidings.

Stations.—When selecting a site for a station, not only should due regard be paid to the proximity and convenience of access to the town or place to be served, but attention should be given to the gradients of the line near the proposed station. If it can possibly be avoided, a station should not be placed in a hollow at the foot of two inclines, as such a position would always entail heavy work starting trains on the ascending gradients, with the risk of sliding back into the station again in unfavourable weather; and for arriving trains there would be increased difficulty in properly controlling the vehicles on the descending gradients so as to bring them to a stand in the event of any sudden stoppage being required. With stations on a summit, having gradients falling in each direction, the starting trains can get away more readily, and the arriving trains have the benefit of the rising gradient to assist them in coming to a stand. Possibly the best selection would be a long length of level, both in the station proper and for a considerable distance on each side; but it is not often that such a combination can be obtained without incurring extra expenditure. The station-yard itself should, however, be on the level, or as nearly so as possible, for the convenience and safety of marshalling or shunting carriages or waggons. No siding should be laid on such a gradient as would render it possible for vehicles to start into motion during high winds. Carriages and waggons having good oil axle-boxes will start themselves on a gradient of 1 in 300 under the influence of a moderately strong breeze, and a slight push will start them on a gradient of 1 in 400.

The number and arrangement of the lines, sidings, platforms, loading banks, and other conveniences of a station, will depend upon the description and amount of traffic to be accommodated. There is a wide range from the simple village station, with its one short siding, to the great city terminus, with its labyrinth of lines and sidings, and its groups of platforms, offices, warehouses, and other accessories. Each station should be laid out with a view to meet the special requirements of the principal traffic likely to arise, whether passenger, timber, coal, stone, cattle, or general merchandise, and ample space should be retained to permit further enlargement and additional sidings at any future time. If provision is not made for the latter in the outset it will certainly lead to large expenditure at some later date. Land adjoining a railway station is quickly appropriated by the public on account of its proximity and convenience for conveyance, and soon covered with store-yards, warehouses, and other buildings, and when any portion of these have to be acquired for station enlargements, they can only be obtained at a large cost, very often ten times as much as the value of the original ground.

When laying out approach roads to goods or passenger stations, whether intermediate or terminal, due importance should be given to the advantage of making them wide, easy in gradient, and fairly straight. A narrow, crooked access to a busy goods yard is a great impediment to the expeditious working of a heavy traffic; and road waggons conveying long pieces of timber or ironwork along such a route, would be very apt to block the roadway and delay the passage of other vehicles. A steep gradient will prevent the carriers taking full loads, and will add to the cost and time of delivery.

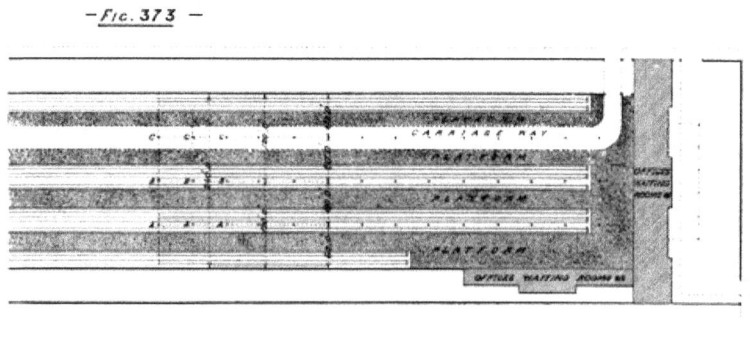

― FIG. 373 ―

An approach road to a large passenger station should be laid out with a long frontage to a wide footpath to enable the numerous intending passengers to alight conveniently from the conveyances which bring them to the station. A portion of the footpath and carriage-way in front of the entrance to the booking-hall should be covered over with a light roof to provide shelter during inclement weather. The footpath should be on the same level as the vestibule or booking-hall, so that the public may pass at once to the ticket-office and their luggage be wheeled on hand-barrows direct to the platform or luggage-room. Every effort should be made to avoid introducing steps from the footpath to the booking-hall, as they check the proper ingress of the passengers, and are very severe on elderly persons and invalids, besides necessitating the dilatory method of carrying each piece of the passengers' luggage by hand. Experience has shown the inconvenience of steps to be so great that in many cases a large expenditure has afterwards been incurred to do away with them, and bring the setting-down footpath to the same level as the booking-hall. For a large station the booking-hall should be spacious and well provided with separate ticket windows for the different classes of passengers and districts of the line; and the access or communication with the platform should be ample and free from obstruction. Small doors and narrow passage-ways check the movements of the passengers and create confusion and delay.

Waiting-rooms for the different classes of passengers, inquiry-offices, luggage-rooms, lavatories, etc., will have to be provided according to the amount of traffic to be accommodated. In large stations it may be necessary to have two or more groups of such rooms to suit the different sets of platforms.

At the most important terminal stations of our home railways it is usual to lay down the main-line arrival platforms with a cab or carriage rank alongside, so that the passengers alighting from the railway carriages have merely to walk across the platforms, and step into the cabs or vehicles waiting to take them and their luggage away from the station. This arrangement is not only a great convenience to the passengers, but expedites the clearing of the platform and the making way for another incoming train. It would not, however, be of any service on continental lines, or other foreign railways, where all arriving luggage must first be taken to the general luggage room, to be examined by the local customs, or *octroi* officers, before being allowed to pass out of the station.

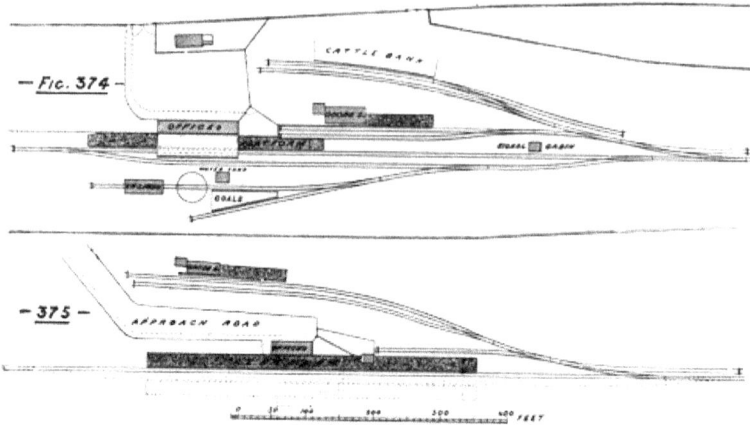

Main-line departure platforms should be of ample width to allow of the free movement of the passengers, ticket examiners, officials, and men wheeling passengers' luggage. The accommodation should not only be sufficient for the normal traffic, but allowance should be made for the large crowds which may assemble for excursion trains during the holiday season or other occasions of national gathering. Additional or local platforms, frequently termed *dock platforms*, may be required for suburban trains, and may be made narrower in width, and without cab ranks, as the passengers using them only travel short distances and rarely have more luggage than they carry in their hands. These dock platforms are generally made available for outgoing as well as incoming trains. The lengths of the main-line or local platforms will be regulated by the number of carriages forming a train.

Fig. 373 is a diagram sketch of a large terminal passenger station, with main and local platforms as above described. It is merely typical to illustrate the principle, and may be multiplied and varied to any extent in the way of lines and platforms. In the sketch the main groups of offices, waiting-rooms, etc., are shown at the end of the station; but they may be equally well placed at the side, as their actual location is principally a question of proximity or convenience of access to some main street or thoroughfare. The lower or platform-level rooms of such a building are mainly devoted to the public for booking-offices, waiting-rooms, refreshment-rooms, lavatories, offices for parcels, telegraph and inquiry, suitable rooms being set apart for lamps, footwarmers, guards, and porters. Above this lower story a range of offices can be built for the use of the principal officers and staff of the different departments of the company.

Fig. 374 is a plan of a small terminal station on a single line of railway, where the passenger traffic is small, and one platform is made to serve alternately both for arrival and departure trains. The booking-hall, waiting-rooms, offices, etc., are laid down parallel to the line of rails, and the approach road and footpath are parallel to the building. The platform roof extends to the outer wall, and provides shelter for the passengers on the platform, and forms a shed for the carriages at night.

Fig. 375 is a sketch of an intermediate or roadside station on a single line of railway. All the offices, waiting-rooms, etc., are on one platform, which serves for trains travelling in either direction. The dotted lines show the additions which would be necessary to make the station a stopping-place for trains working in opposite directions.

Fig. 376 shows an ordinary intermediate or roadside station on a double line of railway, with two passenger platforms, and a connection between them either by subway or over-line footbridge. The principal offices and waiting-rooms are shown on the one side, and only small waiting-rooms, etc., on the other.

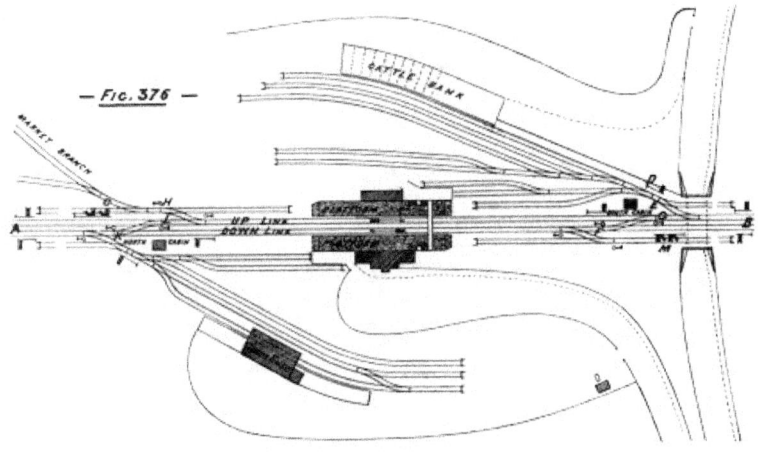

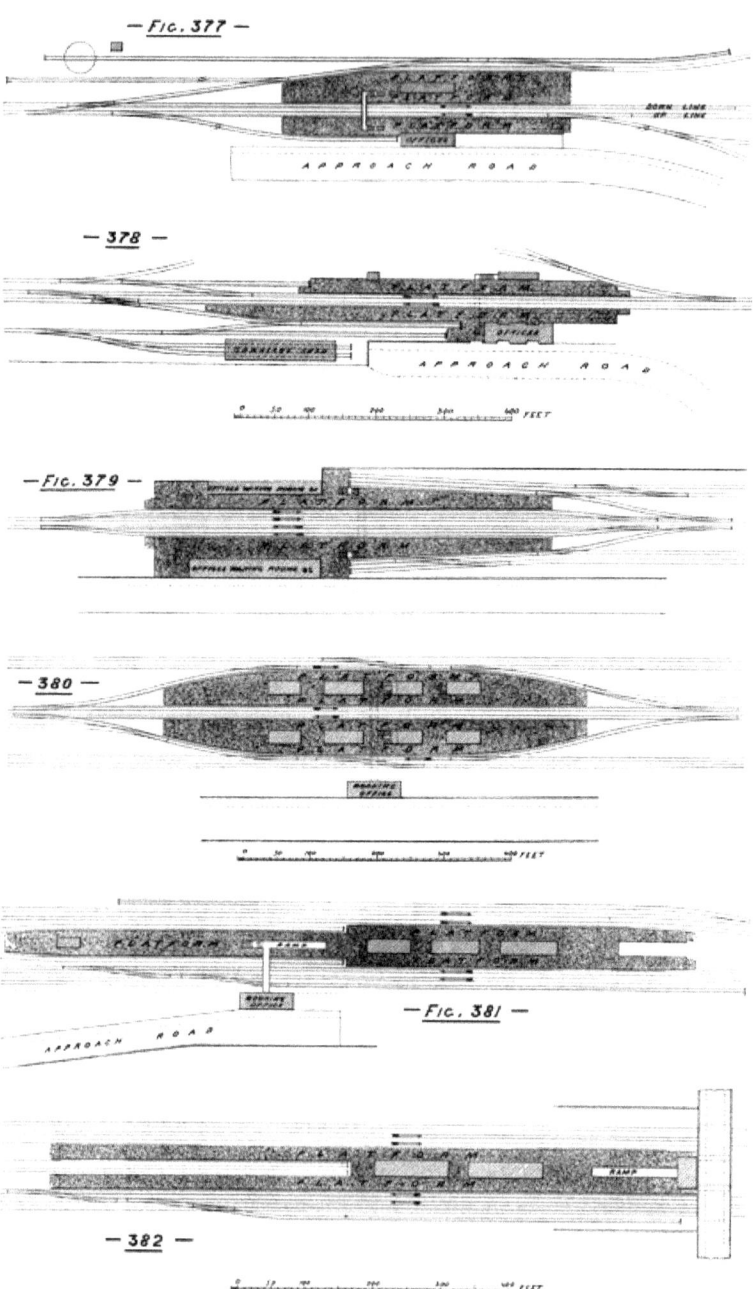

Fig. 377 is a sketch of a double-line intermediate or roadside station at the junction of a small single-line branch railway. Branch-line passengers to and from the main DOWN-line trains merely walk across

276

the platform to get into their respective trains, and those to or from the main UP trains walk across the footbridge or subway to get to the opposite platform.

Fig. 378 is a plan of a double-line roadside station, with two main-line passenger platforms and a dock line and platform for the use of local or branch-line trains. This arrangement is applicable where the actual junction with the main line is at a little distance from the station, but not sufficiently far away to warrant an additional junction station as shown in Fig. 377.

Fig. 379 shows a similar roadside station laid out with a more comprehensive arrangement of dock-lines and platforms. The lines alongside the main passenger platforms are *turn-outs* from the main-line proper, and leave the latter free for the passage of fast through trains or goods trains when an ordinary passenger train is standing alongside the platform. In this way a fast non-stopping train can overtake and be sent forward in advance of a slow passenger train.

Fig. 380 shows a roadside station with two double platforms, the inner lines and platforms being reserved for main-line passenger trains, and the outer lines for branch-line trains. By this arrangement carriages can be quickly transferred from a branch-line train to a main-line train, and *vice versâ*; access from the public road, or from one platform to the other, can be obtained either by subway or over-line footbridge.

Fig. 381 is a sketch plan of an island platform for a double-line roadside station, near which there are junctions with two branch lines. The UP and DOWN main lines run alongside the wide portion of the platform, and the branch lines run into the two dock platforms. The waiting-rooms, refreshment-rooms, etc., are placed in groups on the wide platform, spaces being left between the blocks for the convenience of access from side to side. The booking-office and parcels-office are placed alongside the approach road on the higher level. An over-line footbridge extends from the booking-hall to the dock platforms, terminating with steps on one side and an inclined ramp of 1 in 8 on the other. In carrying out the above plan for a railway on an embankment, the access from the booking-hall to the platform would be provided by a subway instead of an over-line footbridge.

Fig. 382 shows another form of island platform, also arranged for UP and DOWN main-line trains, and two branch-line trains. The access is obtained from a public-road over-line bridge crossing the railway, and the booking-office is placed at the top of an incline, or ramp, leading down to the platform. The dock-line platforms are arranged different to those in the preceding example, with the object of providing longer platforms for the main-line trains. This result, however, is obtained at some little inconvenience to the dock-line trains, as the passengers from one of these must walk round a portion of two platforms to get into the other dock-line train, instead of merely walking across the platform as in Fig. 381.

In some cases of island platforms the total width of the station buildings and platforms is made much greater than indicated in the above sketches, and a wide, easy incline constructed from an over-line public-road bridge, to allow cabs and carriages to come down to a large paved area between the platforms, for the convenience of setting down and taking up the train passengers and their luggage.

The island-platform arrangement possesses many advantages for the exchange of passenger traffic. All the platforms are connected and on one level, and passengers, together with their luggage, can be quickly transferred from one train to another. One set of waiting-rooms, refreshment-rooms, etc., are sufficient, and are available for the passengers of all the four trains. A smaller number of station men are required for the work, as the staff can be more concentrated and better utilized than when there are separate platforms on opposite sides of the line.

The number, size, and arrangement of waiting-rooms and other offices for the public at a large station will depend upon the amount and description of traffic to be dealt with at the particular station under consideration. Where the passenger traffic is to a large extent of a local or short distance character, a moderate amount of waiting-room space may be sufficient, as these local passengers regulate their arrival so as to avoid waiting any great length of time for the trains. An enormous suburban passenger traffic is carried on in many places with a very limited waiting-room accommodation, the frequency of the trains and the routine of the travellers reducing the necessity of such rooms to a minimum. A more ample waiting-room space will be necessary when providing for a large, long journey, or through traffic, and for stations at seaports, as the intending passengers, particularly those landing from steamers, generally reach the station a considerable time before the departure of the trains to take them forward. For this class of traffic it will also be necessary to provide suitable refreshment-rooms. At large terminal stations it is frequently found more convenient for the working of the traffic to have two or more sets of waiting-rooms, etc., separating the local and long-journey passengers, and placing the rooms alongside the corresponding platforms.

Lavatories and conveniences at large stations should be provided on a liberal scale, and fitted up in the most substantial and efficient manner. Not only should they be thoroughly well ventilated, but they should have abundance of light. Nothing tends so much to ensure order and cleanliness in these places as plenty of light.

It will frequently be found that at many of the large important stations there are local surroundings and circumstances of level and foundations, which will to a great extent influence the arrangement of the rooms and offices to be devoted to the public service. No fixed or standard type could be adopted for all cases. Each one will have to be studied out to suit the locality, and the grouping must be made to work in with the best facilities obtainable. In all such cases one of the principal points is to select a convenient position for the booking-hall, easy of access to all persons entering the station premises. On no account should the ticket-office be placed in a position tending to block the thoroughfare on to the platforms. A large number of intending passengers may already be in possession of tickets, and the station arrangements should enable these passengers to proceed at once to the platforms without having to struggle or force their way through crowds of other passengers gathered round the ticket windows. In some instances it is found expedient to provide auxiliary booking-offices for excursion traffic, to be used only on special occasions, thus restricting the principal booking-offices to the ordinary main-line booking.

When laying out small intermediate or roadside stations for either double or single line, or small terminal stations on short branch lines in thinly populated districts, it becomes a question how to provide the requisite statutory accommodation with a minimum amount of building. The following sketches taken from actual examples may be of use for reference.

Fig. 383 shows the smallest size of station building that can very well be constructed to be of any practical service. It comprises an office for the station-master, who has to attend to the tickets, parcels, and telegraph; a waiting-hall with glazed front; a small waiting-room and W.C. for ladies; and a yard with conveniences for gentlemen, coal store, etc. Access to the station is obtained through a gateway in the platform fencing.

Fig. 384 shows a somewhat similar arrangement, but with two additional rooms. The road approach to the station is brought alongside and parallel to the building, and access to the platform is obtained by passing through the booking-hall, which has a glazed front to the line.

Fig. 385 gives the particulars of a building containing rather more accommodation than the two preceding examples.

Fig. 386 shows a small terminal station for a short branch line where there is a moderate tourist traffic during the season. In addition to the regular station accommodation, a refreshment-room is added for the convenience of those passengers who have to drive into the country, or have arrived at the station by road conveyance. The platform roof, which is extended out over the line of rails, as shown on the transverse section, forms a complete covering for the platform, and serves for a carriage-shed at night.

The above sketches merely illustrate types of some small stations suitable for home or colonial lines, and may be built of stone, brick, concrete, iron, or timber. For towns of more importance, the offices and rooms would have to be increased both in number and size. On foreign lines it is customary to provide an office and large hall fitted up with counters for the use of the Local Excise Authorities in the examination of passengers' luggage; and at some stations one or more rooms have to be set apart for the use of the military authorities.

Narrow platforms should always be avoided, especially in front of the offices and waiting-rooms. Nothing tends more to check the proper expeditious working of the traffic than a confined space for the movement of the passengers and of the station staff carrying luggage.

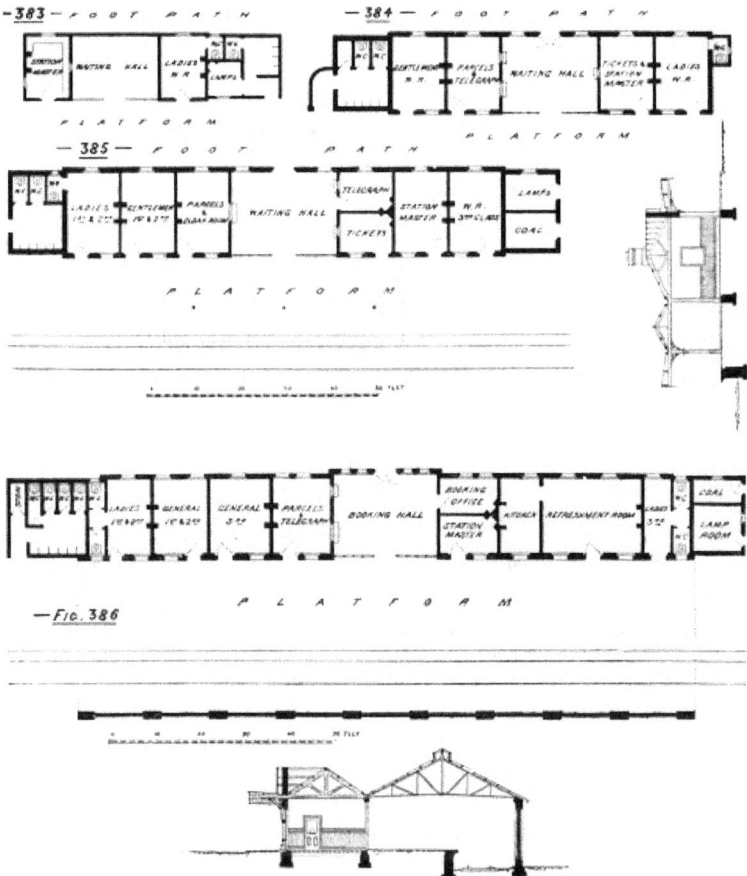
—Fig. 386

In cases where the traffic will warrant the expenditure, it will be found an advantage to construct a light roof or verandah over a portion of the platforms of roadside stations. This covering will provide a convenient shelter for the passengers and their luggage, and prevent the crowding of booking-halls and doorways during inclement weather. In hot countries a verandah or awning of some description on the platforms is an absolute necessity, and those travellers who have had any experience of railways under a tropical sun, will call to mind the celerity with which the passengers seek such welcome shade.

A very important item in the construction of a large terminal station is the roof over the lines and platforms. Wrought-iron and steel can now be obtained in so many convenient sections, and at such moderate prices, that timber-framed roofs, except for very small spans, are now rarely used for railway work. The metallic structure is much lighter in appearance and more durable, besides being less exposed to destruction by fire. The introduction of iron and steel has enabled roofs to be constructed of very much larger spans than would have been prudent to have attempted in timber; at the same time it must be kept in mind that, notwithstanding this increased facility of construction, the cost of a roof per relative area covered increases very rapidly as the span increases. The extent of space to be roofed over in some of our modern terminal stations is so large that the question of roof-spans to be adopted has to be considered very carefully. It has been argued by some that if the area be divided out into small or moderate spans, the presence of the rows of columns for supporting the roof might preclude the possibility of any future re-location of the lines and platforms except by an entire rearrangement of the roof-work. On the other hand, it may also be stated that railway engineers have now obtained such a thorough experience of the necessary relative proportions of platforms and carriage-lines for large stations, as to enable them to lay out these works without any risk of requiring alterations for many years.

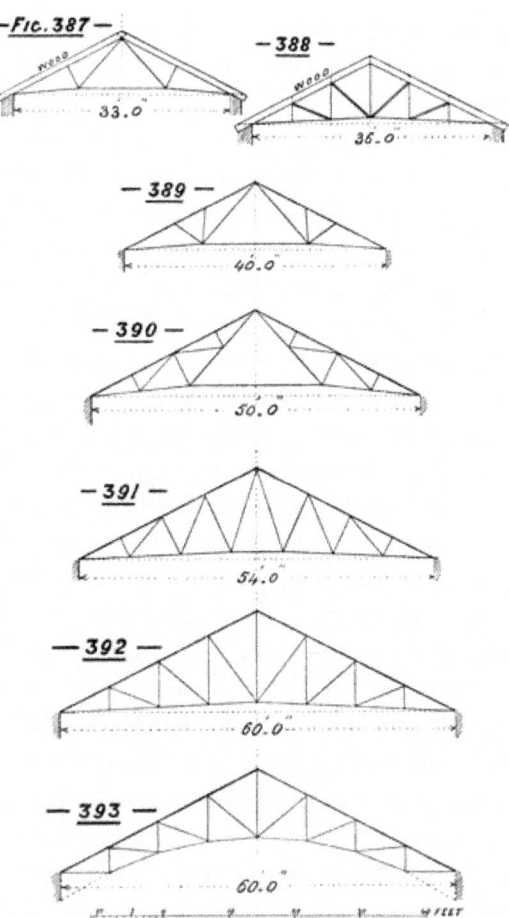

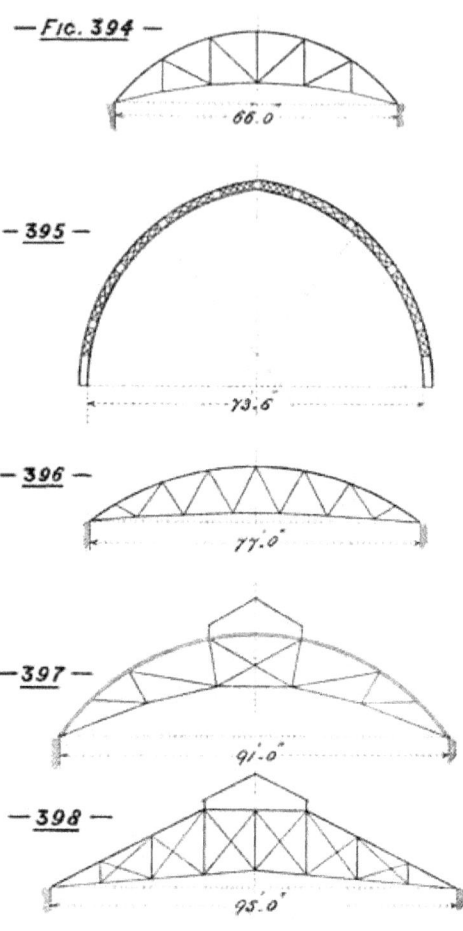

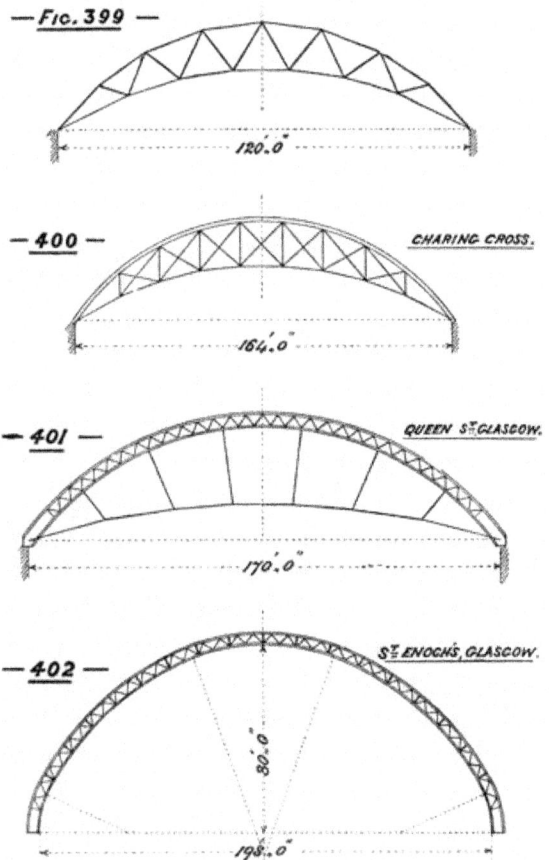

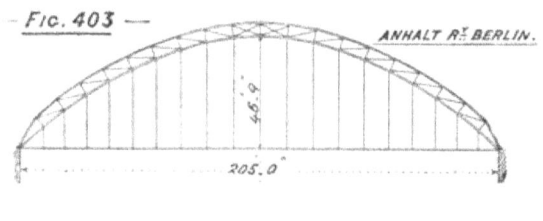

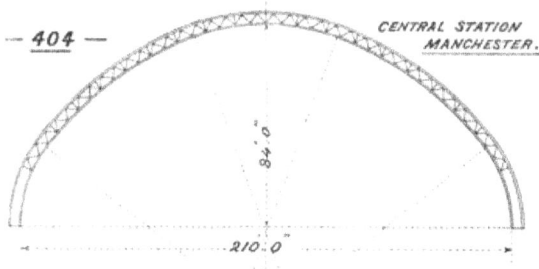

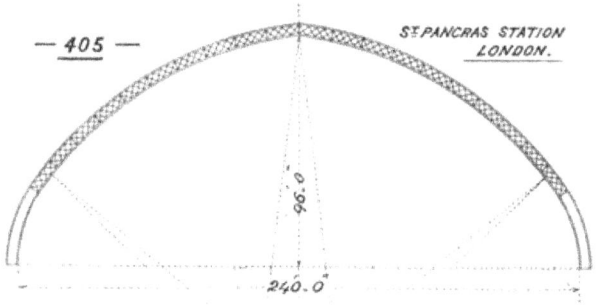

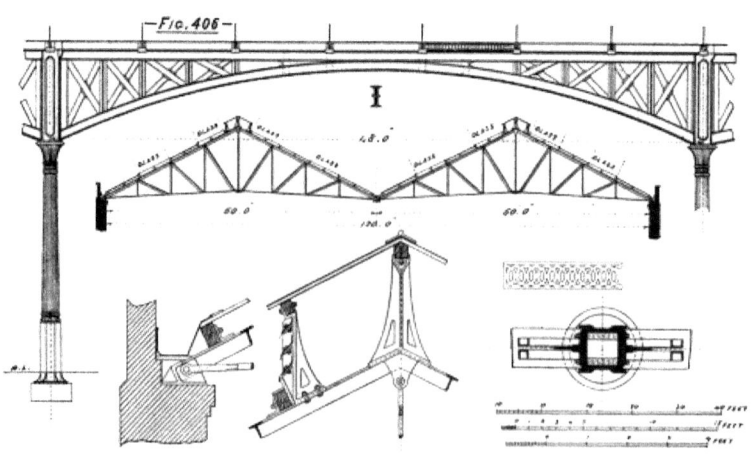

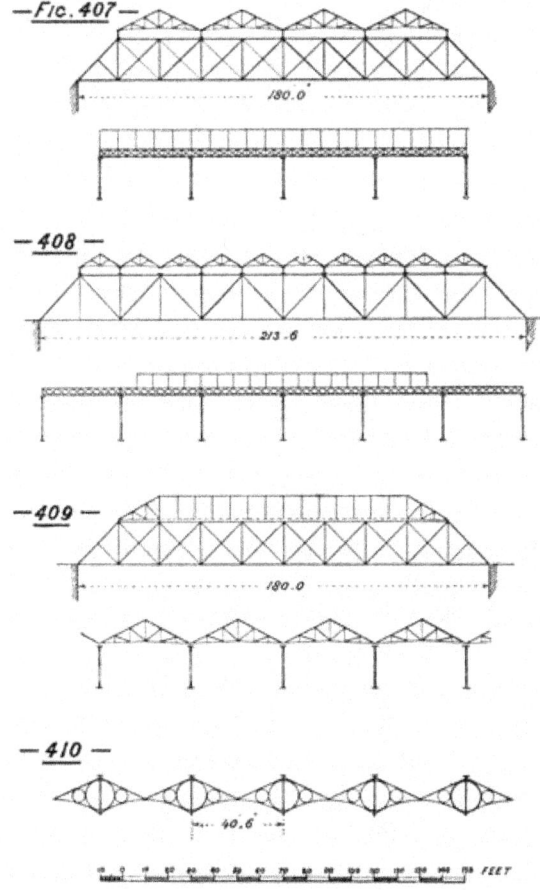

There are so many descriptions of roof-principals used in railway stations that it would be impossible here to introduce more than a few examples. Figs. 387 to 405 illustrate by diagram sketches a series of types taken from actual practice. Fig. 406 gives more in detail the particulars of the roof-principal of 60 feet span, Fig. 392. As will be noted from Fig. 406, the width of 120 feet between the walls is divided into two spans of 60 feet each, the ends of the principals in the centre of the 120 feet being carried on arched wrought-iron girders of 48 feet span, supported on strong ornamental cast-iron columns placed at 48-foot centres. The rain-water from the large centre gutter is taken down inside the columns and conveyed away to drainage pipes laid down for the purpose. The 60-foot principal above described forms a very strong roof, and is light in cost and maintenance. The weight of ironwork, both wrought and cast, in the principals, arched wrought-iron girders, cast-iron columns, centre gutters, etc., is only 0·51 of a ton per square (of 100 square feet) of area covered. For comparison, the weight of ironwork in the roof, Fig. 402, of 198 feet span is 1·42 ton per square of area covered; and of the roof, Fig. 404, of 210 feet span, is 2·07 tons per square.

This increase in weight per square as the spans go on increasing results, not only in a much larger outlay for original construction, but entails also a proportionally heavier expenditure for maintenance and painting. The item of painting alone is an expensive one in all iron-roof work, and must be attended to regularly for the proper protection and appearance of the ironwork. With the smaller spans, the roof-trusses form very convenient supports for painters' scaffolding or planking, but with the very large spans the greater height and the form of the roof-principals render specially designed scaffolding and appliances necessary for the painting and repairs.

Doubtless there is something very attractive about a large span roof, its bold outline stretching from side to side of a wide covered area imparts an imposing effect which cannot be claimed for smaller or more moderate spans; but where roofs are constructed for purely utilitarian purposes it becomes a question worthy of grave consideration whether a series of smaller spans would not provide the same practical benefits as would be obtained from one very large span. Upon referring to the typical sketch of a terminal station, Fig. 373, it will be seen that the total width from inside to inside of main walls is 240 feet. The lines and platforms are so arranged that by placing rows of columns at **A**, **A**, **B**, **B**, and **C**, **C**, the entire width may be divided out into four spans of 60 feet; or, if preferred, a row of columns at **B**, **B** may be adopted, resulting in two spans of 120 feet, or the entire width may be included in one large span of 240 feet. Any one of the three arrangements will provide an effectual roof-covering, and the selection must be decided by the cost or expediency.

Another way to avoid the introduction of large span-roof principals, and to preserve the covered area free from intervening columns, is to erect strong truss-girders extending across at right angles from the main walls. These truss-girders are placed at suitable distances, and carry simple roof-principals of convenient spans. In some cases the roof-principals are placed as shown in Figs. 407 and 408, and in others as in Fig. 409.

In another system the roof-principals are incorporated with the main truss-girders, as in Fig. 410.

With the above type of covering the truss-girders take the place of the arched wrought-iron girders and cast-iron columns, as illustrated in Fig. 406, but will be more costly, as may be gathered from the following brief comparison: Assuming the area to be covered as 480 feet long and 180 feet wide, then the width of 180 feet could be divided into three spans of 60 feet each, or one centre span of 65 feet, and two of 57 feet 6 inches if they would work in more conveniently. With columns at 48-foot centres longitudinally, the three-span arrangement would contain the following:—Twenty cast-iron columns in the two rows, or twenty-two columns if two columns are placed side by side at the extreme end; 960 lineal feet of light arched wrought-iron girders in twenty girders of 48 feet span.

On the other hand, with the truss-girders placed at 40-foot centres to suit roof-principals resting on the tops of girders, as shown in Fig. 409, or to suit the arrangement shown in Fig. 407, there would be twelve heavy truss-girders, each of 180 feet span, making a total length of 2160 lineal feet of deep truss-girder work, exclusive of about another 60 lineal feet, which would be required for the bearings on the side walls.

The successful lighting by day of a large roofed-in station will depend principally upon an appropriate distribution of the glazed portions. With a large span, and the glass skylights placed near the apex, the side lines and platforms will be much less efficiently lighted than those near the centre; and again, if the glazed parts are only at the sides, then the centre portion will be rather in the shade. Where possible it is better to place the glazed portions and slated portions alternately, so as to obtain a more uniform light all over the centre area, somewhat similar to the arrangement shown in Fig. 406.

Roofs over passenger platforms at roadside stations are made in many types, the arrangement depending in a great measure upon the width of platform to be covered. In many of the earlier stations the roof was extended across from side to side, and included the lines of rails as well as the UP and DOWN platforms, a system which was not only costly, but had the disadvantage that the steam and smoke from passing trains remained for some time under the roof before it was thoroughly dispersed. The more modern and more economical plan is to put the roof or shelter over the platforms only, and allow the steam and smoke to pass away into the air. In designing the latter class of roof, the fewer supporting columns the better, so as to diminish as far as possible the obstructions on the platforms. Where the platform is unavoidably narrow, the roof may be carried on curved brackets projecting out from the walls.

Except in tropical countries, where shade is more acceptable than strong light, a liberal amount of glass should be provided in these platform roofs. On many of our home railways they are entirely covered with glass, and the abundance of light is found to be of great assistance in the working of the traffic. Figs. 411 to 420 are sketches of a few out of the many types of small roofs which have been erected over single and island platforms.

Goods-sheds.—The form and dimensions of a goods-shed for any station must be determined by the description and amount of traffic to be transacted at the particular place. With an estimate of the traffic before him, the engineer must consider the internal arrangement of building most suitable for the bulk of the merchandise to be accommodated. The principal object of the shed is to permit of goods being transferred under cover from or to railway trucks or carts without being exposed to the weather, and the transfer will be expedited if the arrangements are made the most convenient for the particular class of merchandise presented.

For some commodities it is considered preferable to unload direct from the railway trucks into carts, or *vice versâ*, and thus have only one handling of the goods. To comply with this method, the cartway must be made almost down to the same level as the rails, to allow the carts or drays to be drawn close up alongside the railway trucks, as shown in Figs. 427 and 428. This type of shed implies a constant supply of carts, so as not to detain the railway trucks, or necessitate the stacking or storing of goods on the low level floor in the way of carting movements.

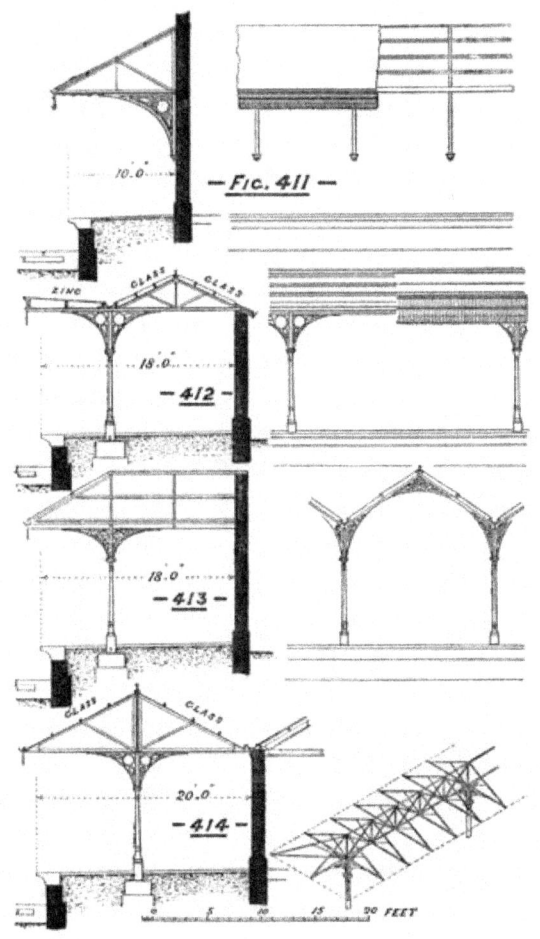

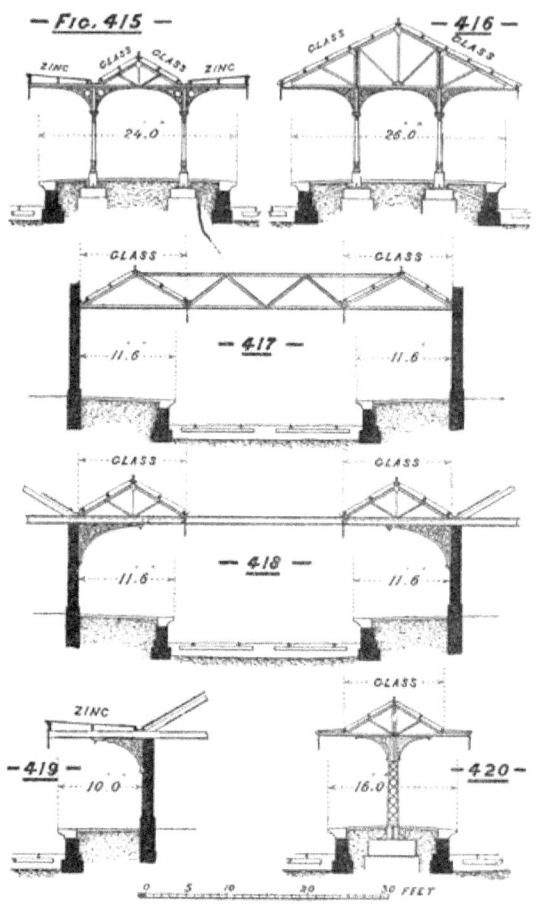

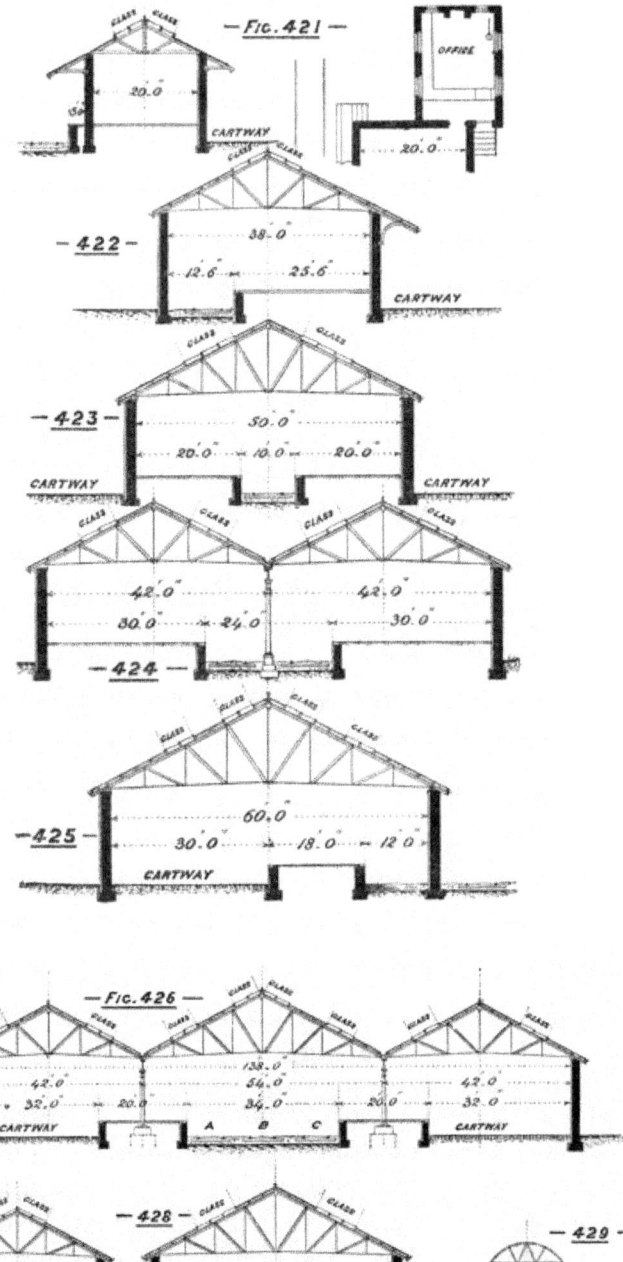

For general merchandise in boxes or bales, a raised loading-bank inside the shed is usually found to be the most convenient arrangement both for loading and unloading. The top of the loading-bank should be a little below the level of the railway-truck floor to give clearance to all truck-doors opening outwards. By means of short portable gangways or landings, the moderate-sized packages are readily transferred to or from the trucks, either by hand or by small two-wheeled trolleys, the heavier pieces being lifted by cranes. The cartway should run parallel to the rails on the opposite side of the loading-bank, and may be either inside or outside the building, according to the importance of the place. When the cartway is inside, the entire front of the loading-bank is available for cart traffic, but this advantage entails a considerable increase in the size and cost of the building. When the cartway is outside, the cart traffic is worked through large doorways placed at suitable distances, and fitted with projecting roofs or awnings to protect the goods during the loading or unloading. At some of these doorways, short docks about 10 feet square, or more, are formed in the loading-bank, into which the carts may be set back fairly into the shed for the greater convenience of the transhipment of the goods by hand or crane power. Where the stacking space is ample, the contents of several railway trucks may be discharged on to the loading-bank without any delay in waiting for carts, and the same railway trucks may be loaded with other goods and dispatched outwards, or may be taken away empty if the loading-bank is reserved for arriving goods only. Where the traffic is large and constant there is an advantage in having separate goods-sheds for the inwards and outwards work.

The following diagram sketches will illustrate some of the many types of goods-sheds in use on railways:—

Fig. 421 shows a shed suitable for general merchandise at a small roadside station. For economy of construction, the line of rails and cartway are both placed outside the building. A small goods-office is built at one end, in which is fixed the pedestal and lever indicator of the cart-weighing machine. The roof is projected outwards over the doorways for the railway trucks and for carts. The railway truck doorways are spaced to correspond to the length of the trucks. A narrow platform, about 3 feet wide, is formed outside the shed alongside the trucks for the convenience of the men when loading or unloading.

Fig. 422 represents a rather larger shed, with the line of rails inside the building and cartway outside. With this type the railway trucks are entirely under cover, and can be unloaded or loaded more conveniently. It has also the additional advantage that the trucks and their contents can be left secure when the shed is locked up at closing time.

Fig. 423 shows a shed with a line of rails down the centre, and a loading-bank on each side, the cartways being outside the building; one loading-bank is for inwards goods, and the other for outwards goods. On the arrival of a loaded railway truck, the door on one side is opened, and the contents unloaded on to one of the loading-banks. The door is then closed, and the opposite door opened for loading from the other loading-bank. By this method a railway truck can be unloaded and loaded again without changing its position.

Fig. 424 represents a shed with two lines of rails down the centre and loading-banks on each side, the cartways being outside the building. One line of rails and corresponding loading-bank is for inwards goods, and the other line of rails and loading-bank for outwards goods. When the railway trucks on the arriving line are unloaded, they are either drawn out of the shed and shunted on to the opposite line to be loaded again, or transferred direct on to the opposite line by turn-tables, or traversers, placed at convenient distances between the columns supporting the roof.

Fig. 425 illustrates a shed in which both the line of rails and cartway are placed inside the building. This is no doubt the most convenient type for transfer of general goods, as all the operations of transhipment are carried on entirely under cover; but it is the most costly, on account of the large building and roof area required.

Fig. 426 shows a large double shed similar in general arrangement to the type represented in Fig. 425, but with three lines of rails down the centre. The line **A** may be used for inwards goods, and **C** for outwards. By means of turn-tables, or traversers, connecting the three lines at convenient distances in the length of the building, the unloaded trucks can be transferred on to the far line, **C**, for loading again, or on to the line **B**, to be drawn away out of the building. The lines **A** and **C** may both be used for inwards traffic, or both for outwards, and the line **B** used for taking away or bringing in empty trucks.

Fig. 427 represents a shed with the line of rails and cartway inside the building, and both very nearly on the same level. This class of shed is often considered the most suitable for fruit, vegetables, and certain light goods which require prompt delivery and careful handling.

Fig. 428 shows a form of shed with a raised loading-bank on one side of a line of rails, and a cartway on the other. With this arrangement the railway trucks may be loaded or unloaded, either from the raised loading-bank or direct from carts and drays drawn up alongside the trucks, according to the description of merchandise presented.

Fig. 429 shows a type of umbrella roof sometimes erected over a narrow loading-bank outside of a goods-shed. It is simple and economical in construction, and provides good accommodation for loading and unloading under cover packages and goods of secondary importance.

The above sketches illustrate some of the many arrangements for goods-sheds, and can be modified and extended in several ways. The leading dimensions, widths of loading-banks, cartways, and gauge of lines, will have to be adjusted to suit circumstances.

Looking at a goods-shed merely as a medium for the convenient transfer of merchandise between the railway and the roadway, the inference is soon drawn that the removal of the goods into trucks or carts should be effected as speedily as possible, otherwise a large extent of shed-room will be required for carrying on a moderate amount of work. Every effort should be made to clear the goods from the loading-bank as soon as they have been properly unloaded and checked. Any laxity in this respect will cause an outcry for increased accommodation, which a little more energy and careful organization would have prevented.

Timber plank floors are generally preferred for inside loading-banks. Inside cartways should be formed either of granite setts or wooden-block paving; the latter is better, being less noisy, and, if occasionally sprinkled with sand, will afford a good foothold for the horses. A macadamized roadway under cover is never satisfactory, as it is always dry, and never binds together into a compact even surface. Sliding or rolling doors are the best for goods-sheds, as they are more out of the way, and under better control during high winds.

Cranes of appropriate strengths, and worked by hand or other motive-power, should be distributed in suitable positions throughout the shed. They should be placed so that they can, when required, lift direct out of a railway truck on the one side, and deposit into a cart or dray on the opposite side of the loading-bank.

Goods-sheds may be built of stone, brick, iron, or timber, or a combination of all of them. Where the requirements are well proved, and the traffic certain, it is better to build a substantial permanent structure. Iron sheds, with sides and roofs of galvanized corrugated iron sheets, will last for many years if not made of too light materials. There are many cases where it is more prudent to put up a goods-shed in timber than to incur the cost of one of more permanent character. Where the traffic is uncertain, or the foundations bad, or out in undeveloped districts abroad, a building of timber will serve the purpose for a number of years, or until the period of probation has passed, and the actual requirements are accurately ascertained. In a timber-built shed, the decay usually commences about the ground line, but if the nature of the soil will permit of the construction of a small dwarf foundation wall of masonry or concrete up to about nine inches above the ground line, the life of the building will be prolonged for several years.

The best method of admitting daylight into a goods-shed is from the roof, and a liberal extent of roof-glazing should be provided for the full length of the building, and so distributed as to be well over the loading-banks. In tropical countries the amount of roof light must be reduced, on account of the great glare from the sunlight.

An ample supply of artificial light will be necessary when working after dark or during the night. In some instances the goods-sheds in large and important business centres have one or more upper storys, in which goods are warehoused pending the owners' instructions, the goods being transferred between the loading-banks and upper floors by lifts or cranes.

A proper supply of weighing machines for carts, drays, railway trucks, and packages on the loading-banks will be necessary to facilitate the checking of the goods.

There is always a large proportion of traffic which can be dealt with outside the goods-sheds, either on loading-banks or cartways alongside the sidings. Outside loading-banks should be of good width, with approach roads of easy gradient. In tropical countries a light shed, open on all sides, is frequently erected over a portion of these outside banks, to protect the goods and workmen from the heat of the sun. Fixed cranes or travelling cranes will be required for lifting the large packages, heavy castings, and logs of timber. Where there is a large cattle traffic, separate sidings, loading-banks, and approach roads should be set apart for the purpose, with suitable water-troughs and cleansing appliances. Horses can be unloaded at any loading-bank, but for the more valuable class of animals and for carriages it is usual to construct a special horse and carriage dock, as shown in Fig. 430, the carriages being wheeled off the end of the carriage truck, as indicated in the section. Cartways alongside the sidings are very convenient for unloading coals, stone, bricks, sand, lime, and many other materials which have to be passed out of the trucks in small quantities at a time. To encourage and facilitate traffic at roadside stations, traders are frequently allowed to stack or store large supplies of some of the above materials on ground set apart for the purpose near some convenient siding, the stock being disposed of in detail to suit the local requirements. Coal-drops are sometimes adopted where there is a large trade in that commodity. They are constructed by carrying the line of rails on strong balks of timber or small girders placed across the top of walled-in coal-yards or divided areas. The coal is thrown out of the trucks, and falls a depth of 15 or 20 feet into the yard below. In consequence of the height from rail-level to ground a large tonnage can be piled up, and stored in a small area, and the unloading of the trucks effected very rapidly, particularly so where special trucks with opening floors or hinged bottoms are used for the purpose. In many cases capacious roofed-in sheds are built for storing coals, lime, cement, grain, or other materials liable to deterioration from the weather. These sheds are built alongside a siding; the contents of the trucks are unloaded or thrown into the sheds through doors spaced to correspond to the railway-truck doors, and are carted away through doorways on the opposite side.

It is customary to place *buffer-stops* of some form at the termination of dead-end sidings in a station, to bring to a stand such carriages or waggons as may be approaching with too much speed to be stopped without the interposition of some substantial barrier.

Figs. 430, 431, 432, and 433 are sketches of some of the many kinds of buffer-stops, and will explain themselves. In Fig. 430 the buffer-stop is made of flange rails, and is shown as fitted in a carriage-dock with wrought-iron plate landing, **A**, and plate-iron hinged flaps, **B**, **B**. The latter are turned over, and rest on the floor of the carriage-truck, to form a pathway when taking on or off a vehicle.

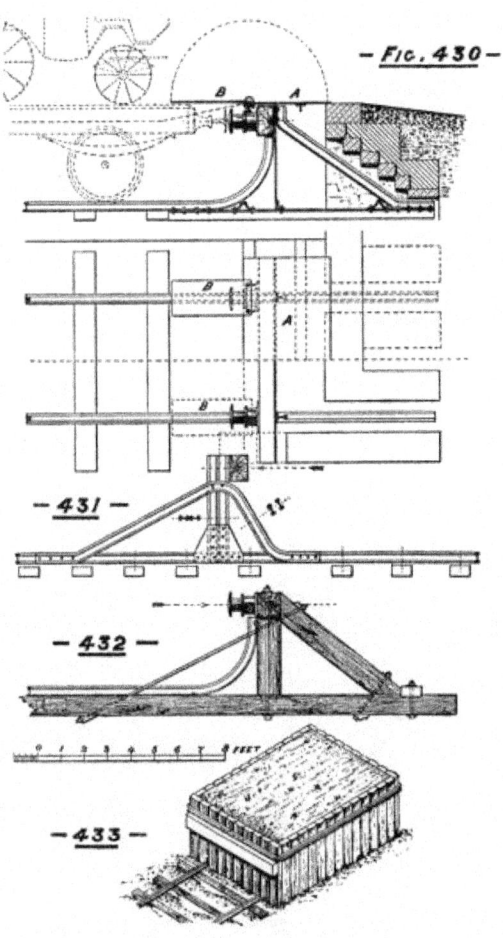

Fig. 431 shows a buffer-stop made of double-head or bull-head rails; and Fig. 432 is a buffer-stop made of heavy timbers.

Fig. 433 shows a very simple buffer-stop frequently adopted for sidings where there is not much traffic. It is made of good old sleepers bound together with old double-head rails, and the interior filled with earth or clay.

In addition to the buildings alluded to in the foregoing description, the engineer has to design and construct very many others in connection with railways. These will include large running-sheds for stabling working locomotives; sheds for housing carriages; workshops for building and repairing engines, carriages, and waggons; foundries; large stores for materials; offices; dwelling-houses; mess-rooms, etc.; many of them involving questions of difficult foundations, and nearly all of them requiring special strength and stability to meet the heavy weights and vibrations to which they are subjected.

CHAPTER V.

Sorting-sidings—Turn-tables—Traversers—Water-Tanks and Water-Columns.

Sorting-sidings.—On many important long main lines it is necessary to establish special independent sidings for sorting or arranging waggons of merchandise and minerals. Where there are only two lines of rails to serve for the UP and DOWN service of a heavy passenger and goods traffic, it is imperative to restrict those lines as much as possible to the actual transit of trains, and not to block them by unnecessary occupation for shunting purposes. A goods train running a long distance collects waggons from many roadside stations, and at some of them several waggons will be taken on, to be forwarded to various and widely distant destinations. The accumulated train comprises waggons which must be divided out into groups, to be passed on either to distant sections of the same railway system, or on to neighbouring lines. To avoid interruption to the train-working, and the delay of complicated shunting operations at the roadside stations, the waggons are attached just as they are dicked up, and the work of sorting is allowed to stand over until the train arrives at the place assigned for the purpose. A site for sorting-sidings is generally selected where the ground and gradient are favourable, and where ample room can be obtained for a large number of short parallel lines, with space for future extensions. The arrangement that naturally suggests itself is that of a series of fan-shaped sidings leading out of main shunting lines, separate from the main-traffic lines. In some cases the sorting-sidings are laid down with dead-ends, as in Fig. 434, and in others they are made as through sidings, connecting at both ends with shunting lines and main-traffic lines, as in Fig. 435. Each of the sidings is usually made sufficiently long to hold a complete train of sorted waggons, and the number of them will depend upon the number of sections to be served, and the amount of waggons to be sorted. Sometimes the sidings are laid with a slight falling gradient leading away from the main shunting lines, to facilitate the running out of the waggons into the respective sidings.

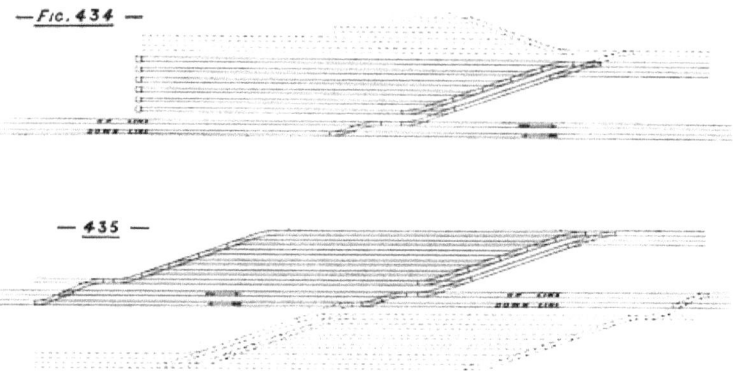

An arriving goods train is first drawn out of the main-traffic lines into one of the shunting lines, and then handed over to the staff of men in charge of the sorting operations, who at once mark the waggons according to the number or designation of the particular siding into which they have to be placed. A suitable engine is generally set apart for this work, and in a very short time the entire train is divided out by one or more waggons at a time, and distributed into the various sidings, representing different sections of the line, or groups to be handed over to neighbouring railways. When one of these sorting-sidings contains a full complement of waggons, an engine is attached, and the train despatched to its destination, leaving the siding clear for another set of waggons. Where the trains to be sorted are very numerous, two or more shunting-engines may be engaged working at the same time on distinct sets of shunting lines and sidings. Sometimes it may be expedient to have one lot of sorting-sidings leading off the UP line, and another lot leading off the DOWN line, to meet the requirements of trains coming and going in both directions. With sidings well laid out, and fitted with ample facilities, a well-organized staff can carry out a very large amount of work both expeditiously and economically. There are several of these sorting-sidings stations in operation, where from one thousand to two thousand waggons are sorted and marshalled into trains every twenty-four hours.

The above diagram sketches merely illustrate the general principle of the sorting-sidings, and may be modified and enlarged in many ways to suit the traffic requirements and local surroundings.

Turn-tables.—Turn-tables revolving on fixed centres are made of various sizes according to their use for engines, carriages, or waggons. The carrying-beams may be made of cast-iron, wrought-iron, or steel, but the latter material is the most suitable for tables of more than 20 feet diameter. For small turn-tables, cast-iron beams will serve very well, for although more liable to fracture, they will not suffer so much from rust and oxidization as wrought-iron or steel.

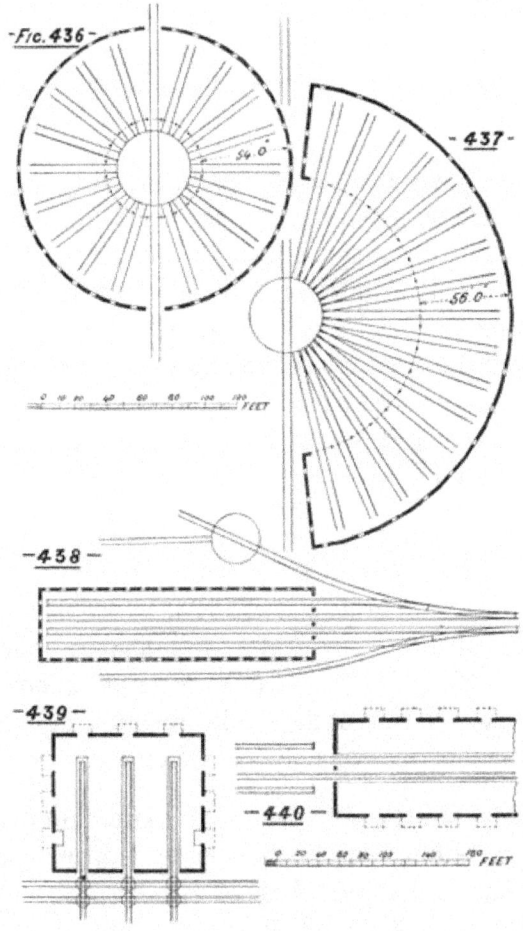

Opinions as to the most convenient position and use of turn-tables have undergone a considerable modification during the past twenty or twenty-five years. Circular and semi-circular running-sheds for engines, as in Figs. 436 and 437, are not so often adopted now as formerly. Although compact and accessible in theory, they possess the one great drawback that when the turn-table in the centre becomes deranged by wear or accident, none of the engines on the standing-lines inside the building can be taken out until the turn-table is again put into working order. A stock of from twenty to thirty engines might thus be put entirely out of the service for a day or more. This objection is considered to be of so serious a nature that running-sheds are now almost always constructed of rectangular form, of which Fig. 438 is a type.

With this description of shed, the lines of rails are laid down parallel to one another, and the engine turn-table is placed on a line separate and distinct from those lines forming connections with the shed.

Where there is a large goods traffic, an endeavour is generally made to so lay down the goods-sheds and approach lines and sidings, that the full complement of waggons may be shunted in or out of the shed at one operation. This arrangement, which dispenses with turn-tables altogether, admits of the ready removal of a central or far-end waggon, without the necessity of taking out so many others in front one by one over the turn-table. At the same time, there are large numbers of these waggon turn-tables in use, and there are many cases where access to side sheds or detached stores can only be obtained by turn-tables.

A goods-shed and lines laid down with turn-tables, as in Fig. 439, will always be more tedious and costly to work than one laid down with direct through lines, as in Fig. 440. Should either of the turn-tables shown on Fig. 439 get out of order and become incapable of turning, then the entire side of the shed controlled by that table will be rendered useless until the defect be remedied.

Engine turn-tables are rarely made with more than one road on the top. The most modern types generally consist of two strong wrought-iron or steel-plate girders well braced together and securely attached to a middle framework which rests on and revolves round a centre-piece fixed on a solid foundation To the ends of the girders are attached large roller wheels which travel round a solid iron or steel roller-path laid down along the circumference. These modern turn-tables are generally worked on the balancing principle, by bringing the engine and tender to a stand in such a position on the rails that the greater portion of the weight is thrown on to the cup-shaped steel centre, so that a small force applied to the long outrigged hand-levers at the ends is sufficient to turn one of the heaviest locomotives. Figs. 441 and 442 give sketch plan and section of one of these steel-plate girder turn-tables, which has few parts, and very little to get out of order. The end rollers guide the table when making any portion of a revolution, and carry such part of the weight as may not be taken up by the centre. A recess is shown in side wall to facilitate the inspection of end rollers. In the earlier forms of engine turn-tables, the revolving movement was effected by attaching to the upper portion of the girders a strong winch, which acted upon gearing fixed either to the end rollers, or direct on to a toothed ring forming part of the roller-path. In cases where the engine turn-table was in constant use, as in connection with a large running-shed, the winch was sometimes driven by a small steam-engine to expedite the movement.

The great increase in the lengths and weights of modern locomotives has necessitated the removal of many of the old small turn-tables, and replacing them with others of 45 or 50 feet, or more, in diameter.

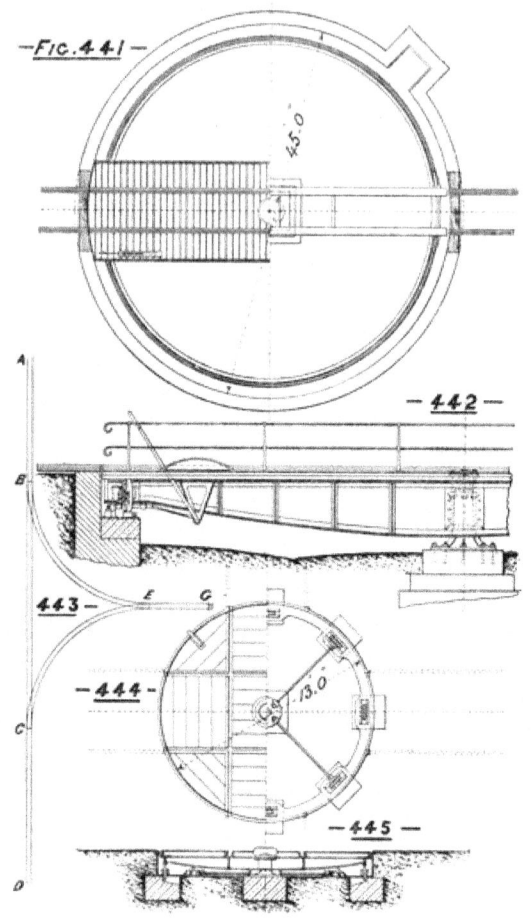

An engine turn-table is a costly item in railway requirements, not only in the girder-work, but in the large amount of building in the side walls and centre pier, and an effort is always made to avoid the outlay unless the table can be placed where it may be of permanent use. In the construction of foreign railways, and in our colonies, where the lines are opened in sections as the work goes forward, the temporary arrangement shown in Fig. 443 is frequently used instead of an engine turn-table. The sketch will almost explain itself. On the main line, **A**, **B**, **C**, **D**, switches are placed at **B** and **C**, from which turn out curved lines, uniting at the switches **E**. An engine proceeding from **A**, and passing round the curve **B**, **E**, **G**, then round curve **G**, **E**, **C**, and back along main line, **D**, **C**, **B**, **A**, will be turned round as efficiently as on a turn-table. The writer has used this arrangement abroad with great advantage. It involves very little work or expense beyond laying down the permanent way, and so soon as the temporary terminus of the line has been advanced further ahead, the rails and sleepers can be lifted and used again elsewhere.

Figs. 444 and 445 give sketch plan and section of a waggon turn-table which has been largely adopted. The centre should be securely fixed on a solid foundation of masonry, brickwork, or concrete. The deep outer cast-iron ring is made in segments, properly fitted and bolted together, and fastened down to the foundation course. The stop-checks are cast on to this outer ring. Two roads, at right angles to each other, are laid on the turn-table, so that waggons to or from the goods-shed have only to make one quarter turn of the table. The top is generally covered with either chequered iron plates or timber to give good foothold for the men and horses which have to pass over in moving the waggons. If properly balanced, the table is easily turned by men pushing at the opposite corners of the waggon, or by a horse and tail-rope, or by hydraulic power through a capstan. In many cases of bad or soft foundations these small turn-tables are erected on a strong framework of creosoted timber.

Carriage turn-tables are now very rarely used. With the old short four-wheeled carriages the moderate-size turn-table was convenient for transferring an extra carriage to or from a spare carriage-line alongside the making-up train at a platform, but modern carriages are now so much longer, some of them twice the length, or more, than formerly, that nothing less than an engine turn-table would be large enough for them. Sometimes a carriage traverser is used for this station work, but much more frequently these long carriages are shunted on or off the making-up train by simply running them in or out through the nearest switches and cross-over road.

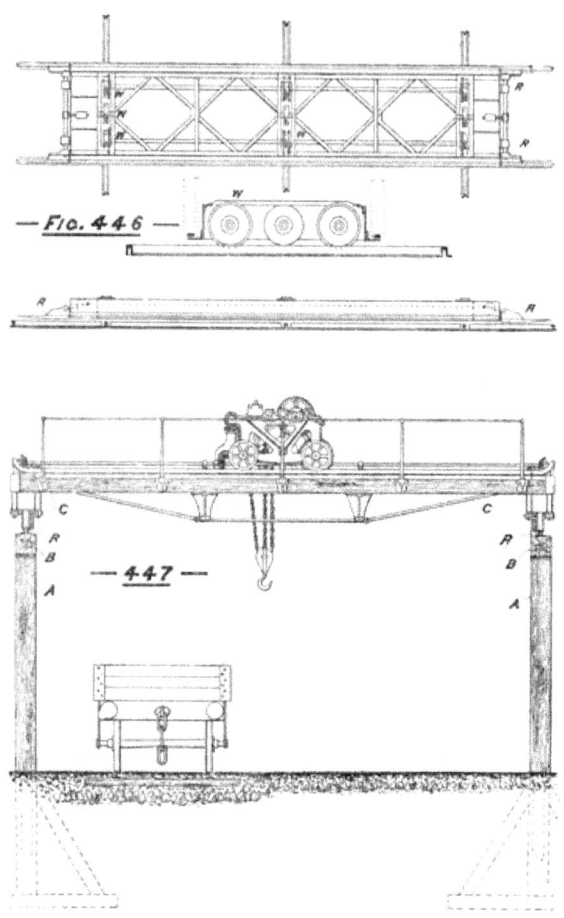

Fig. 446 is a sketch of a carriage-traverser, of length to suit an ordinary six-wheeled carriage. The length, however, may be extended to take on a bogie carriage or any other long carriage. The framing is made of wrought-iron or steel, well braced together. The carrying wheels, **W, W**, run upon rails laid at right angles to the running-line or siding, and the carriage is moved on to or off the traverser by means of the hinged ramps shown at **R, R**. A carriage, once on the traverser, may be moved across one or several lines of running road, according to the extent of traverser line laid down; and this appliance is very suitable for large terminal stations and carriage-repair shops. It will be observed that the operations of the turn-table and the traverser are quite distinct. With the former a vehicle can be transferred from one line to another, and also turned completely round; but with the traverser the vehicles are simply moved in a parallel direction, from one line to another, and when it is necessary to turn or change a vehicle end for end, as in the case of a mail-bag-catching apparatus van or a special saloon, then resort must be had to a turn-table.

Cranes.—A large portion of the merchandise conveyed on railways must be lifted into or out of the trucks by cranes. The position, description, and capacity of these will depend upon the materials to be handled. Large slow-working powerful cranes will be necessary for raising heavy castings, large logs of timber, or massive blocks of stone; while the small quick-acting cranes will be more suitable for dealing with the lighter packages, casks, and bales.

Fig. 447 shows a gantry or overhead crane, used for lifting heavy weights out of an ordinary road-waggon, carrying them a short distance, and then depositing them in a railway truck, or *vice versâ*. Double-flanged rollers, attached to the ends of the platform **C, C**, run upon the rails **R, R**, which are fixed on the top of the beams **B, B**, secured to the verticals **A, A**. The working length of the gantry is only limited by the number of the verticals, and this, being the fixed portion of the work, may be extended out to any distance required. The travelling or carrying girders of the platform **C, C** may be made of wrought-iron, steel, or timber. They must be strongly framed and braced together as a platform to carry the lifting machinery and weight lifted, and have convenient gearing for effecting the transverse or side-to-side movement, as well as a horizontal movement along the line of rails on top of the verticals. Where the fixed portion of the gantry is of considerable length, two or more travelling platforms can be used. In the sketch given above, the entire gantry is shown as made of timber, but iron or steel can be equally well adopted, and continuous masonry or brickwork walls may be built to serve as verticals.

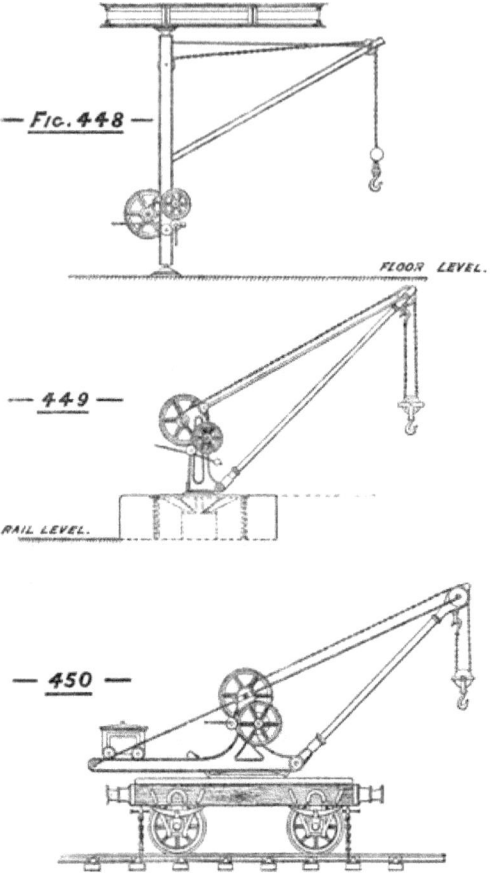

Fig. 448 is a sketch of a small handy crane for warehouse work; it is quick in action, and restricted to weights not exceeding twelve hundredweight. This form of crane may be strengthened to lift still greater loads, but in doing so the additional size of the parts, and the corresponding extra labour in working, detract from its efficiency as a quick-acting crane for light weights.

Fig. 449 shows an ordinary fixed three-ton jib crane, a very convenient size for general station work. The centre pillar is fixed into a bed of masonry or a solid block of concrete. The jib is of wrought-iron or steel, those materials being so much more reliable than timber, and very little more expensive. This crane must be fixed so that in one direction the jib may command the centre of a railway truck, while in the other it can conveniently raise the packages to or from the carts or loading-bank alongside. In the sketch the crane is shown as placed on the loading-bank, but it may be placed on the same level as the rails if preferred. Cranes of this type and strength are frequently found necessary for the inside work of goods-sheds, where packages of considerable weight have to be handled. A very similar class of jib-crane is constantly made for lifting weights of five or ten tons or more, the different parts being made stronger and heavier to correspond to the weights to be raised.

Fig. 450 shows a five-ton travelling crane. Although more costly, it has the advantage over a fixed crane that it can be moved about from place to place. It is mounted on a very strong waggon framework, and provided with springs and spring buffers. Instead of moving round a long deep centre, the jib of the travelling-crane is arranged to work round a bevelled metal roller-path laid down on the platform of the waggon, and has a heavy counterweight loaded to correspond to its capacity. Before commencing to lift any weight strong oak blocks or filling pieces are inserted between the tops of the axle-boxes and the under side of main beams of waggon, to relieve the springs of the pressure which would arise from the weight lifted. From the four corners of the waggon are suspended chains carrying gripping-hooks to be attached or clipped round the rails. These gripping-hooks, when firmly secured to the rails, prevent the crane from tilting over, as the weight of the waggon and also of the rails and sleepers are brought into play to counteract any tendency to throw the crane off its proper balance. With the larger size travelling cranes, capable of lifting ten or fifteen tons or more, outriggers of joist or I-iron, moving in slides, are run out at right angles on either side, and can be loaded with bars of iron or other weights to form a counterpoise.

A medium-sized travelling-crane is a most useful appliance about a railway station; it has a much greater range of utility than a fixed crane, but it is not always appreciated as it should be. It merely requires a line of rails laid down parallel to the rails of siding, and may be placed either on the same level as the siding, or on the level of the loading-bank. Being laid flush with the roadway, the rails do not present any obstacle to the passage of carts or movement of merchandise. As one waggon on the siding is loaded or unloaded, the crane can be moved along its own line of rails, and be put to work at another without the necessity of moving or drawing out any of the railway waggons on the siding. Five, ten, or twenty, or more railway waggons can be dealt with in this way, according to the length of crane-line laid down. The crane can also be readily removed to another part of the station-yard, or to another station along the line. For stations with an intermittent or spasmodic traffic in heavy timber, large blocks of stone, or other unwieldy articles, a travelling-crane is particularly suitable, as it will meet all the wants so far as the lifting is concerned, and when the rush of traffic is over, it can be easily transferred to some other sphere of usefulness. The crane-siding itself is never very costly, as the rails are generally old rails taken out of the main line, and laid on good second-hand sleepers. They have little to do, and merely form a track for the moving crane.

Fig. 451 is a sketch of an ordinary Goliath crane constructed of timber. The general arrangement and capabilities of this crane are somewhat similar to those of the gantry shown in Fig. 447. Both of them are designed to lift heavy weights, and move them sideways into, or out of, ordinary road waggons, but the methods of application are different. In the gantry the verticals are permanently fixed, whereas in the Goliath the verticals and overhead girders are all attached and braced together, forming a complete framework which is carried by double flanged rollers running on the lines of rails **R**, **R**. The winches or gearing for lifting the weights, or slinging them sideways, or for propelling the crane forward on the rails, are attached to the verticals as shown, and are worked from the ground-level instead of the overhead platform, as indicated in the gantry. As each Goliath crane is complete in itself, there is nothing to prevent two or three of them working at the same time on a long length of crane-line.

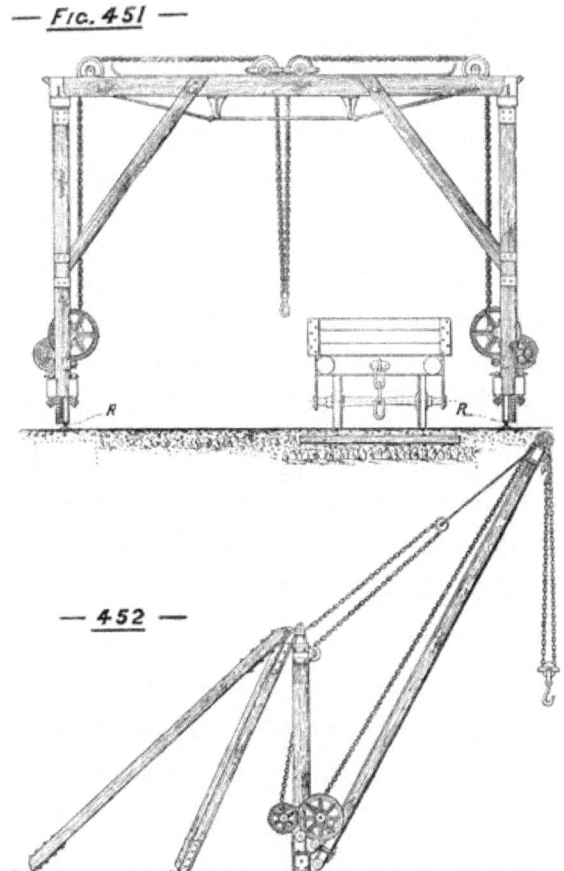

Fig. 452 shows an ordinary derrick crane, which, on account of the large and varying sweep of the jib, is found very convenient for certain classes of work. It occupies a considerable amount of room, and its adoption is therefore limited to situations where space is of secondary importance.

All the cranes described above are shown as worked by hand-power, but they may be worked by steam, hydraulic machinery, or electricity. Manual power will be the most economical where the use of a crane is only occasional, but it would be too slow and costly where there is constant heavy work.

Water-tanks.—A supply of good water forms an important item in railway working, and ample provision must be made at all principal stations for the requirements of engines and general station purposes. According to the locality, the water may either be procured from the main of some established waterworks company, or be pumped from a well, or forced up from a stream by a ram, or brought down by gravitation in pipes from a spring or stream at a distance. Water thus obtained is conducted into tanks placed at a height of 18 or 20 feet, or more, above the level of the rails, and forms a storage supply from which deliveries can be made at a fair pressure and in large volume. The tanks may be made of cast-iron, wrought-iron, or steel, or even of wood. In the great timber-producing countries abroad, water-tanks, some of them of large capacity, are very frequently made of wood, the circular or half-cask form being preferred; but at home, and on European lines generally, wooden tanks are rarely used except for temporary purposes. Cast-iron being less liable to deterioration from rust than wrought-iron or steel, is much used for water-tanks.

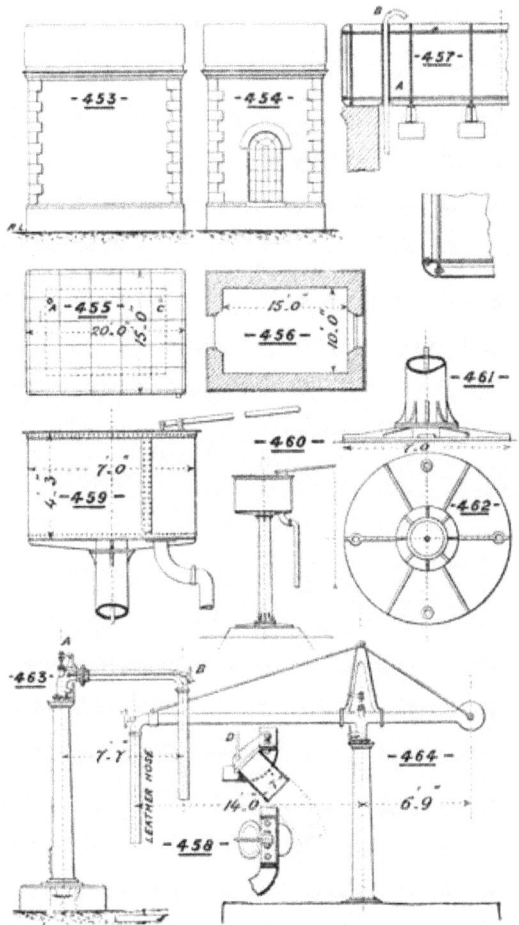

Figs. 453 to 457 are sketches of a medium-sized cast-iron water-tank, to hold about 7800 gallons. The size may be varied both in length, width, and depth, without in any great measure altering the type. The lower portion, or tank-house, may be of stone, brick, wood, or iron framework, and may be utilized as a pump-room, store, or lamp-room. In the sketch given a row of cast-iron girders are placed across the top of the walls of the tank-house, to carry the tank, the plate-joints of the latter being made to coincide with the centre lines of the girders. The lower and upper edges of the tank-plates are shown curved in section, the former for appearance and facility of cleaning, and the latter to check the tendency of the water rippling or splashing over the sides when disturbed during high winds. The large pipe, **A**, is securely bolted at the bottom of the tank, and forms a shield or funnel through which the supply pipe, **B**, passes upwards into the tank. **C** is an overflow, or waste pipe, to carry away any surplus which may find its way into the tank after the water has risen to its fixed maximum height. All the contact surfaces of the cast-iron tank-plates must be accurately chipped or planed, and fitted to ensure water-tight joints. Stay-rods must be placed at frequent intervals, connecting the vertical or outer plates to the horizontal or floor plates. When required to hold more than 20,000 gallons, it is better to make the tank in two parts, by placing a permanent plate partition across the middle, in reality making two separate tanks, which can be connected or disconnected at will. The double tank arrangement gives additional strength, and possesses the advantage that the one tank can be emptied and cleaned out while the other remains in service.

Water-tanks constructed of wrought-iron or steel plates are usually made circular in form, with vertical sides. The floor-plates must be either carried on small girders, as in the cast-iron tank, or be strengthened internally with angle-irons, tee-irons, and tie-rods. The rivetting must be well done, all joints sound and watertight. This class of tank must be kept well painted, or oxidization will take place very rapidly. The arrangement of inlet, waste-pipe, and delivery pipe may be the same as for the cast-iron tank. Although frequently seen abroad, these circular wrought-iron tanks are not often adopted at home. By many the appearance of the circular tank is considered inferior to one of neat rectangular shape, and the form of the round tower does not lend itself so conveniently for use as a pump-room or store.

There may be no practical difficulty in constructing a large circular wooden vat or water-tank, but there cannot be any great actual economy, except in those countries where suitable timber is very cheap, and iron very dear. The wooden tank must be made of selected materials, and by skilled workmen; but however carefully constructed it cannot be expected to last so long as an iron tank. In many parts of the United States of America there are excellent examples of the circular wooden tank, strongly put together, and covered with a light ornamental roof. Numbers of these wooden tanks have been erected there in places where the cost of carriage alone of an iron tank would have been a serious item, and where suitable timber was fortunately close at hand.

In cases where engines are watered direct from a water-tank, a simple delivery-valve, as shown in the sketch (Fig. 458), will answer the purpose. This valve has to be pulled open by the chain and lever, **D**, and when released falls with its own weight, and is kept closed by the pressure of the water above. The delivery-pipe should not be less than 7 or 8 inches in diameter, to accelerate the filling of the tenders. Where water has to be delivered to engines at two or more places in a station-yard, and the supply derived from the same principal tank, the result may be obtained either by laying down 7 or 8-inch main pipes from the principal tank to separate water-columns, or by erecting two or more pedestal water-tanks, similar to Figs. 459 to 462, each of which holds a little more than the average quantity for one tender, and can be fed from the principal tank by a comparatively small pipe of 3 or 4 inches in diameter. It is simply a question of expense—whether it is cheaper to lay down a long length of 7 or 8-inch main pipe and ordinary water-columns, or to adopt the small pipes and pedestal tanks.

Figs. 459 to 462 are sketches of a medium-sized pedestal water-tank to hold 1200 gallons. The supporting column must have a very wide base, bolted down to a solid foundation. The tank itself, made circular in plan, is generally constructed of light plates of wrought-iron or steel, the lower portion or floor of tank being very securely attached to the vertical column. Notwithstanding their top-heavy appearance, these pedestal tanks can be made very firm and steady if enough width be given to the base-plate, and the tank properly fixed to the column. Water is led into these pedestal tanks by a small pipe passing up inside the supporting column, and the delivery may be effected by a simple valve, as explained for Fig. 458.

Fig 463 shows one type of water column for watering engines. The wide base-plate is bolted down on to a foundation of stonework, brickwork, or concrete, and the main supply pipe (not less than 7 or 8 inches in diameter) is carried up inside the column, and connected with the screw valve, **A**, which regulates the delivery to the tenders. The curved top, which forms the outlet, and carries a leather hose, works on a swivel joint, and can be swung round, either to the right or left, for convenience of supplying engines on one or two standing-lines. The delivery valve can be opened or closed by the small hand-wheel **B**, which is conveniently accessible to the man on the tender.

On the above sketch (Fig. 463) the water column is shown placed at an ordinary normal distance from the rails; but in cases where there is considerable space between the two lines of rails, or where a platform intervenes, the swinging arm may be extended out to the necessary length, and counterbalanced as shown in Fig. 464.

CHAPTER VI.

Comparative Weights of some Types of Modern Locomotives.

Weights of Locomotive Engines.—The demand for higher speeds of passenger trains, with more conveniences, luxuries, and consequent increased weights in the carriages, has naturally led to greatly increased power and weight of the locomotives devoted to the passenger service. Although these engine weights have so largely increased during the past twenty-five years, there is nothing to indicate that they have yet reached the maximum. The tendency is still to increase, and will doubtless continue, so long as the permanent way can be made to sustain such enormous rolling loads. Locomotives for goods trains have also increased in power and size, but perhaps not in the same proportion as those for the passenger service. There is not the same disposition to expedite the transit of goods and minerals, which do not deteriorate during a long journey. Perishable articles, such as fish, fruit, and milk, are usually conveyed by passenger trains, or trains set apart specially for the purpose.

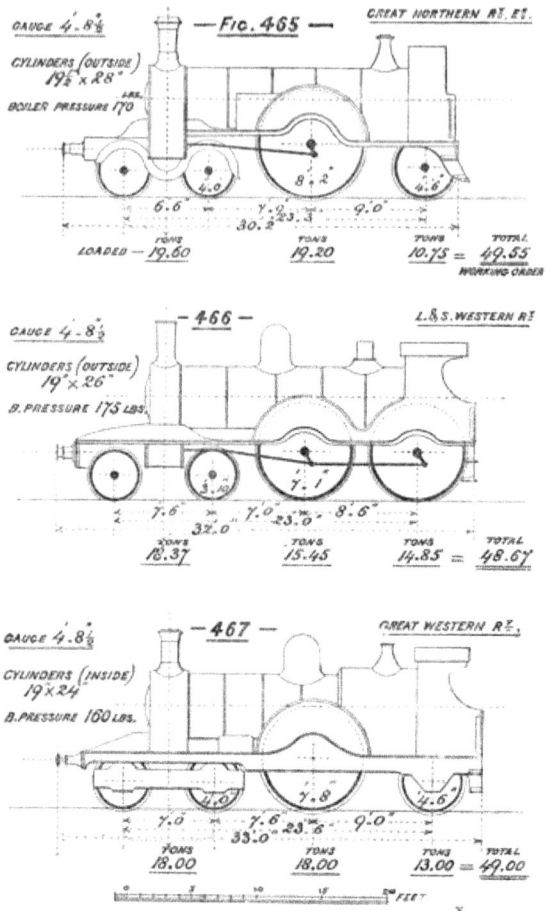

The heavier engine doubtless possesses greater tractive power, but apart from the question of tractive power is the all-important one of steadiness and safety on the rails. A locomotive passing round a curve, even at a moderate velocity, produces disturbances in proportion to the capability of the machine to adapt itself to the altered position, and if both the engine and permanent way are constructed so as to be almost unyielding, then destructive wear and tear and increased risk of derailment must ensue. The adoption of the four-wheel bogie truck to the locomotives on our home and continental lines—although very slow in coming—has contributed greatly to their improvement, enabling the weight to be distributed over a longer, yet more flexible wheel-base, affording greater facility and comparative safety in traversing curves; and rolling, or passing over the rails, with as little injurious effect to them as possible. It is strange to find that the four-wheel bogie truck, originally designed in England in the early days of the railway era, should for so long have met with so little favour on this side of the Atlantic. The Americans, at all times prompt to recognize any appropriate mechanical arrangement, adopted the bogie truck upon its first introduction into the States. They have worked out many improvements in the details, and upon the thousands of locomotives on their vast network of railways, the bogie truck, in one form or another, has been universally adopted from the beginning.

On our home and continental lines, the modern express locomotive, with a four-wheel bogie truck in front, is a much longer vehicle than its predecessors, and its total weight is distributed over a greater wheel-base; but the actual weight placed upon the driving-wheels, or on the coupled wheels, is now very much in excess of former practice, and must be taken into consideration when working out the details of girders and cross-girders of under-line bridges. Numbers of girder bridges have had to be taken down and replaced with stronger structures, not for reasons of wear or decay, but simply because they were incapable of carrying with safety the modern heavy rolling loads. Present experience points out the expediency of providing in all new under-line bridges a liberal margin of strength to meet future developments.

Figs. 465 to 479 are diagram sketches of a few modern types of locomotives, giving leading dimensions and weights, and may be found useful for reference when working out the necessary strengths of the various portions of bridge-work. Upon comparing some of the principal particulars with those of the earlier class, it will be noted that in many of the modern types the piston area has been doubled, the boiler-pressure doubled, and the weight of the engine doubled also.

The engines shown in Figs. 465 and 467 have great weights placed on the single driving-wheels, and should only be used where there is a very strong permanent way. With the four-wheel coupled engines, the weight for adhesion can be distributed between the driving and trailing wheels.

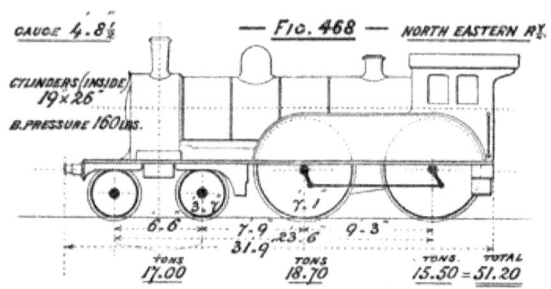

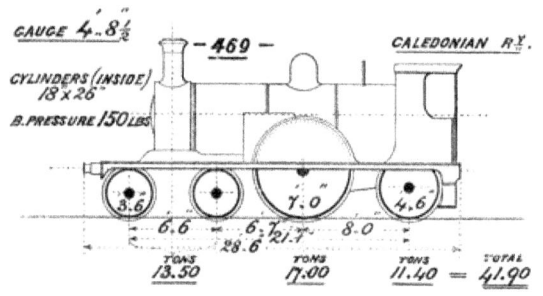

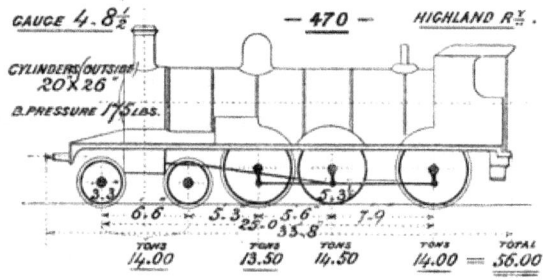

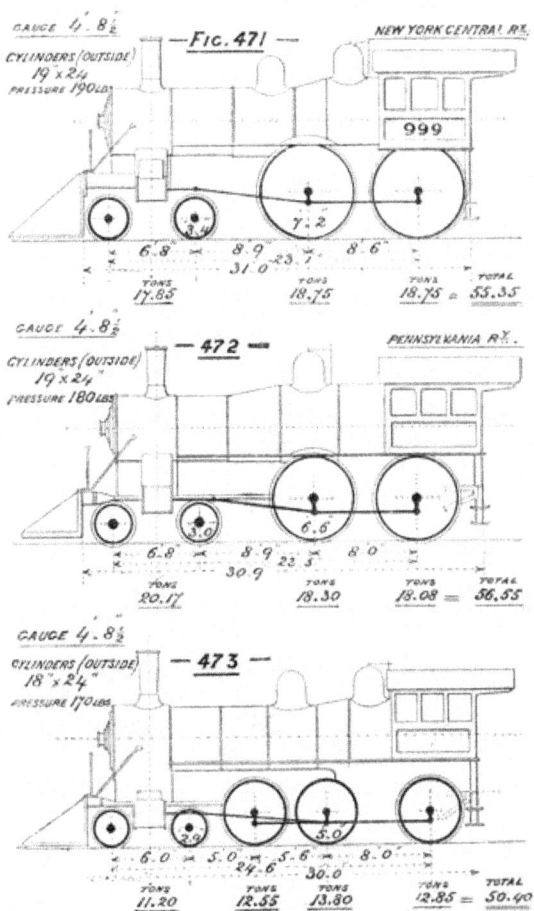

Fig. 473 represents a very excellent type of American engine which has been extensively adopted in the United States for many years. The six coupled wheels distribute the weight over a fairly long wheel-base, retaining their united weight for adhesion. The four-wheel bogie truck in front forms a valuable path-finder to the engine, both for passing round curves or on straight line. This class of engine is very serviceable for various kinds of traffic, and is particularly suitable for lines where the rails and fastenings are comparatively light. In the example shown, the flanges are turned off the centre pair of coupled wheels; but for lines where the curves are of small radius, the flanges may be turned off the leading pair of coupled wheels, instead of the centre pair, to reduce the length of rigid wheel-base. This type of engine has latterly been introduced on various European and foreign railways, and recently on the Highland Railway of Scotland, as shown in Fig. 470. The writer has had engines of this class under his charge abroad, and found them to be most useful for heavy passenger and goods-train service. They run very steadily, are easy on the permanent way, and light in repairs. As they become better known they will be more appreciated, and will doubtless before long supersede in many cases the rigid six-wheel-coupled goods engine. The principal objection of any importance that can be raised against them is that on many lines the present engine turn-tables are too small for such long engines; but it would be far more economical in the long run to enlarge a few turn-tables than to continue the adoption of rigid engines which from their form and arrangement tend to unnecessary wear to themselves and the permanent way.

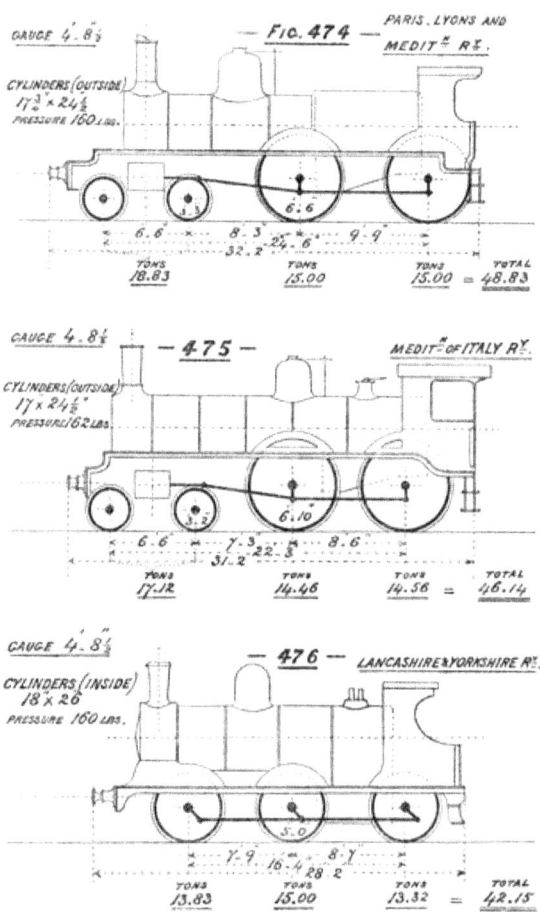

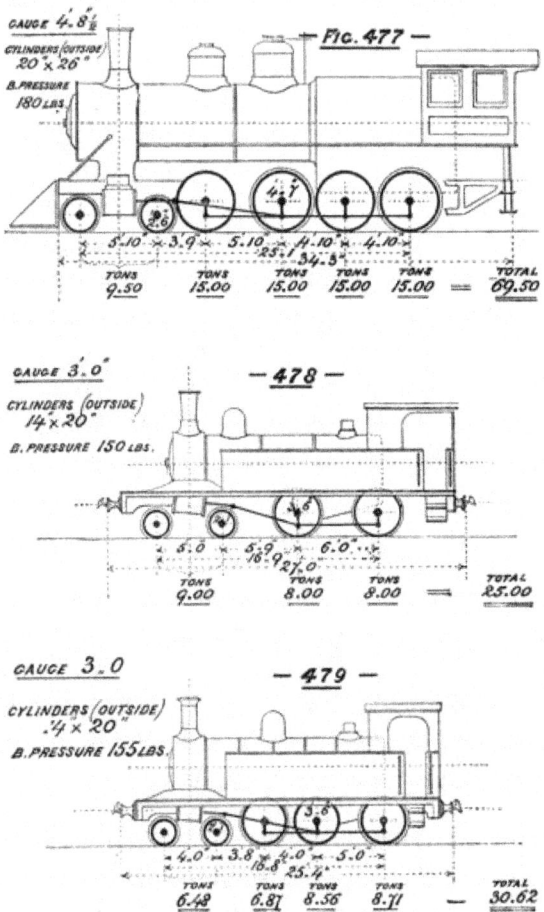

Fig. 476 shows an average sample of the ordinary six-wheel-coupled goods engine in use on so many of our home railways. Where the curves are easy and the permanent way strong, the drawback of the long rigid wheel-base may not be so apparent; but for a line abounding in sharp curves, perhaps no more destructive machine could possibly be devised than the ordinary six-wheel-coupled goods engine. Without any flexibility, forced along with great power, and too often driven at unnecessary high speeds, engines of this type have too small a margin of safety when traversing the curved portions of the road. A slight unevenness in the rails, or a sharp flange on the wheel may supply all that is wanting to cause the engine to leave the track, and the probability that such risks are more common than is supposed, is far from satisfactory. The great weight of the engine doubtless tends to keep it on the track, but the rapid wear of the tyres, and of the inside of the rail-heads clearly demonstrate the enormous amount of friction and abrasion that takes place.

Fig. 477 represents a type of eight-wheel-coupled engine designed for hauling passenger or goods trains over long lengths of heavy mountain inclines. The engine is a large one in every way, and of great total weight, but the weight is distributed over a long wheel-base and without imposing a greater tonnage per pair of wheels than is done in some of the smaller and less powerful engines. The flanges are turned off the leading pair and third pair of coupled wheels reducing the rigid wheel-base for curves to 9 feet 8 inches. The four-wheel bogie truck in front carries only a moderate weight, being so close to the coupled wheels. Engines of this description require a strong permanent way, as there is a total weight of 60 tons on the four pairs of coupled wheels standing on a wheel-base of 15 feet 6 inches.

Figs. 478 and 479 are types of tank-engines in use on some of the narrow-gauge (3 feet) railways. In general design they are somewhat similar to the modern class of engine on main lines of 4 feet 8½ inches gauge, with four-wheel bogie truck in front, and four wheels or six wheels coupled, but with all the parts and weights smaller, to suit the narrow gauge and lighter permanent way.

The extended use of the bogie truck is an admission of its advantage over the fixed-wheel arrangement, both for distribution of weight and facility in passing round curves; but although it is now so largely adopted for engines and carriages on our home and continental railways, it is somewhat of an anomaly to find it so very rarely used for tenders. In the United States all the locomotive tenders—and many of them of very large size and weight—are carried on two four-wheel bogie trucks, and traverse the curves as easily as the engines. On this side of the Atlantic, the prevailing custom is to mount the tender on six rigid wheels; and as many of these tenders weigh as much as from 35 to 40 tons in working order, and have a rigid wheel-base of 15 feet, it will be seen at a glance that much unnecessary friction and wear and tear would be avoided by substituting two four-wheel bogie trucks for the fixed wheels.

CHAPTER VII.

Signals—Interlocking—Block Telegraph and Electric Train Staff Instruments.

Signals.—Railway tradition alleges that on one of the early lines opened for passenger traffic, the precautions for public safety were considered to have been fulfilled by providing a man on horseback to ride along the track between the rails in the front of the locomotive engine, to give warning to persons strolling on the line, and to check the advance of the train when necessary. A very short experience of this method of working proved that the full capabilities of the locomotive could not be obtained from a restricted speed of seven or eight miles an hour, and a more comprehensive system of signalling had to be devised. By fencing in on both sides of the line, the public were prevented from making a general highway or promenade along the railway, and the problem was reduced therefore to the signalling for the trains alone.

Flags of different colours, held by flagmen stationed at suitable places, answered the purpose for a while, or so long as the authorized running speed did not prevent the train being brought to a stand after sighting a flag warning the engine-driver to stop. As speeds were increased, a longer or more distant view of signals became imperative, and tall posts, or semaphore signals, were introduced. Well-defined blades or discs placed on high posts were easily worked from the ground-level, and could be seen for long distances, thus enabling the trains to be controlled or brought to a stand before reaching the signal. The efficiency of the principle once recognized, improvements and additions were made from time to time, until we have the simple acting tall semaphore signal so universally in use at the present time. The position of the signal arms or blades in the daylight, and the colours shown by the lamps at night, form the code of signals for the proper working of the train service; and as the signal arms and lamps are both worked simultaneously by the same gearing, it is only necessary to light the lamps to put the signals in complete condition for night-working. For some years, when the traffic was small, with trains at low speeds and at considerable intervals, one double-arm semaphore signal-post at a station was made to serve for all purposes; but as the train service became more frequent and more rapid, it was found that another semaphore or tall post signal, was necessary to give warning to the engine-driver some distance back before reaching the station or *home signal*. More particularly was this necessary at those stations where it was not intended that every train should stop. This new signal, called the *distant signal*, very soon came into general use. It was placed at distances varying from 400 to 800 yards away back from the station or home signal, and was worked by a long strained wire extending from the distant signal to a ground-lever placed near the home signal, the levers for these distant signals and home signals being thus near together and under the control of one man. More recently it was found necessary to introduce another important wire-worked signal called a *starting signal*, which is placed at the outgoing or departure end of the passenger platforms, lines, or station sidings, to prevent any train or engine starting or proceeding on its journey until such starting signal is lowered to indicate that the line is clear.

These simple, independent, hand-worked semaphore signals did good service for many years, but being independent and in no way physically connected with one another at junctions, or stations, or with the switches they were intended to control, it was quite possible for mistakes to arise where everything depended upon the accuracy and prompt decision of the signalman. The possibility that such mistakes could occur, and the certainty that they actually did occur, and too often with most disastrous results, led gradually to the grouping and interlocking of a large number of signal levers and switch levers together in one signal cabin. The advantages of the concentrating and interlocking of signals and switches are twofold. In the first place, one man in the signal cabin can work and control the levers for a large number of switches and signals, where formerly several men were required to be located at various places in the station-yard; but the second, and by far the most important advantage, is that with proper interlocking arrangements it is practically impossible to give conflicting signals.

With a modern interlocking frame, and assuming the normal position of all the signals to be at *danger*, then before a signal can be lowered for an approaching engine or train all the switches and corresponding signals, from any lines or sidings connecting with the line to be signalled *clear* must first be set so as to prevent any engine or train coming out of such connecting lines or switches on to the line to be made clear. In a similar manner, before the points and signals can be set to permit an engine or train to pass from a siding on to the main line all the necessary signals must first be set to *danger* to prevent the approach in either direction of any engine or train on the main line about to be occupied. The mechanical arrangements of the interlocking frame are so exact and complete as to effectually prevent any but the proper combination being made. An untrained or inexperienced signalman might inadvertently attempt to pull over a wrong lever, only to find it securely locked and immovable under the control of other levers. The proper sequence of levers must be made, and the accurately adjusted mechanism automatically prevents mistakes which formerly occurred with the old hand-worked signals from the oversight or confusion of the signalman.

The interlocked switches or points are worked from the signal-cabin by light wrought-iron tubing (termed rodding) or channel-shaped iron bars supported on fixed iron rollers, and the signals by galvanized wires running over light pulleys. Modern signals are always weighted at the signal-post, so that in the event of the breaking of the pulling-wire they will fly back to their normal position of danger.

The facility and precision secured by the interlocking machinery enabled other valuable accessories to be introduced for the more complete signalling and protection of train-working. Amongst these may be mentioned the facing-point bolt-lock and rocking-bar, signal-detectors at points, and throw-off or trap points.

With the old-fashioned hand-worked switches the man standing alongside could see whether the sliding-rails were properly closed, and also when the last vehicle of the train had passed over them; but when important main-line-facing switches or points are worked by rodding from a signal-cabin some distance away, it is necessary to have some reliable means to ensure that the sliding-rails are actually brought close home, and also to prevent the switches being moved again until the entire train has passed over them. A set of switches may be carefully made and work well, but it is quite possible for some fracture or obstruction, to intervene and prevent them closing properly. If a train or engine were passing through them in a trailing direction, as indicated in Fig. 345, the wheels would most probably force the sliding-rail home, and no disturbance would arise. If, however, the train were coming in the opposite or facing direction, the chances are that some of the wheels would take one road and some the other, and cause a derailment. The same casualty would occur if the switches were moved during the passage of the train.

To guard against the above contingencies, the facing-point bolt-lock and rocking-bar have been introduced. The system is applied in various forms, but the arrangement shown in Fig. 480 will explain the principle generally.

A strong casting, **A**, is securely bolted to the top of the sleeper carrying the chairs on which rest the point ends of the sliding-rails. This casting has an internal groove or chamber formed for its entire length from **C** to **D**, as indicated by the dotted lines, and in which slides the locking-bolt **B**. The point ends of the switch or sliding-rails are connected by the transverse rod **E**, which is forged into a vertical bar form for that portion of its length, which passes through the opening, **F**, prepared for it in the casting **A**. In this vertical bar a hole or slot is cut to correspond to the exact size of the locking-bolt **B**, and at a distance to suit the sliding-rails when pulled over to their properly closed position. This locking-bolt, **B**, will not pass through the hole in the vertical bar until the sliding-rails are quite close home, and when once through the hole the sliding-rails cannot be moved until the locking-bar is withdrawn. In some cases two holes or slots are cut in the vertical bar to enable the points to be bolt-locked for both directions.

The rocking-bar is designed to prevent the withdrawal of the locking-bolt before all the vehicles have passed over the points.

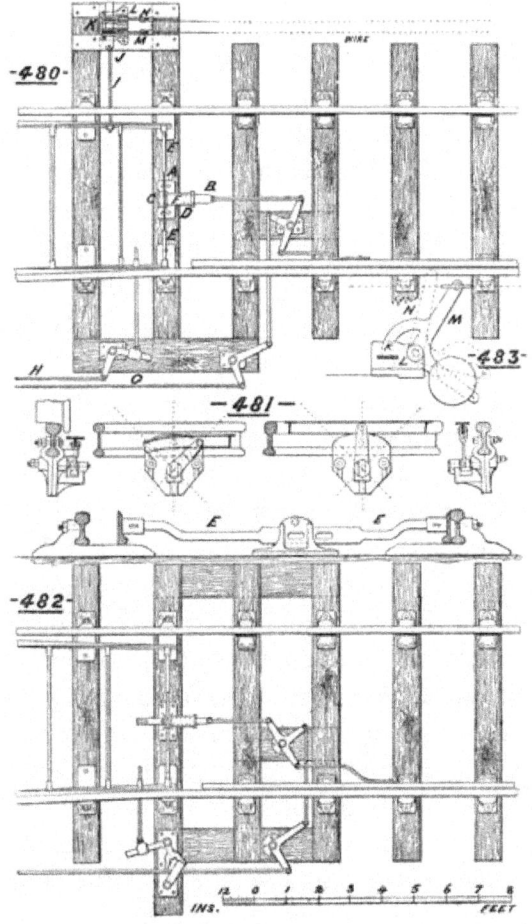

This rocking-bar consists of an angle iron or tee-iron bar of a length equal to the longest wheel-base of the rolling-stock, and is carried on short pivoted arms working in cast-iron or wrought-iron brackets secured to the rails as shown in Fig. 481. The pivoted arms have a movement backward or forward, and when at either the one or the other extremity, the upper surface of the rocking-bar is sufficiently below the top of the rail to be well clear of the flange of any passing wheel; but while changing from the one to the other position, and when the pivoted arms are vertical, or at half-stroke, the upper surface of the rocking-bar is about level with the top of the rail, and right in the pathway of the wheel-flange. It is evident, therefore, that when the pivoted arms are set in the forward or backward position, and one of the wheels of a train or vehicle has passed on to the rail over the rocking-bar, the latter cannot be changed or raised and pulled over to the opposite extremity so long as any one of the wheels of the train or vehicles remain over the rocking-bar.

As the same ground-crank which pulls over the pivoted arms from backward to forward also withdraws the locking-bolt **B**, the latter is thus held securely in the hole or slot of the transverse rod, **E**, until all the wheels of the train have passed off the rocking-bar. The operation of changing the points from one road to another is very simple. By means of the rodding **G**, worked by a lever in the signal-cabin, the locking-bolt **B** is first withdrawn from the slot; the points are then pulled over into the reverse position by the rodding **H**, and the locking-bolt **B** is again set back into one of the slots by the rodding **G**. Sometimes, for economy, the points, bolt-lock, and rocking-bar, are all three worked by one lever in the signal-cabin, and one set of rodding on the ground, as shown in Fig. 482; but the arrangement is neither so perfect nor so secure as that shown in Fig. 480. Where there are two sets of rodding and gearing, the failure or breaking of either of them prevents the complete combination being made, and indicates at once to the signalman that something is wrong; but when there is only one set of rodding a breakage may occur without giving any tangible evidence to the signalman of the defect, and he may proceed to pull over his signal lever in ignorance that the points have not been properly made and bolted. To avoid an accident taking place from the failure of either rodding or gearing, the signal-detector has been devised, so as to prevent the possibility of pulling over the signal wire until the points and locking-bar are both in their proper positions.

The signal-detector is applied in several forms; the one shown in Figs. 480 and 483 will explain the principle on which its efficacy depends. A transverse rod, **I**, attached to the sliding-rail, extends out beyond the rails, and is formed into a flat bar or plate, **J**, sliding through the guide-holes **K**, **K** in the casting **L**. Short upright levers, **M** and **N**, work on trunnions fixed in the casting, and to **M** and **N** are attached the wires leading from the signal-cabin and continuing on to the signal-posts, as shown in elevation in Fig. 483. Two slots are cut in the plate **J** to receive the curved arms of the levers **M** and **N** when they are drawn downwards to pull off the corresponding signals.

Neither of the levers, **M** or **N**, can be drawn over unless there is a slot immediately under the curved arm into which it can enter. When there is solid plate under a curved arm, the short lever cannot be pulled over, and the signal therefore remains at danger. The slots in the plate **J** are spaced so that one will be brought into position for one of the curved arms, when the points are close home for the main line, and the other slot for the other curved arm, when the points are set for the branch line or siding. The two slots cannot be under the two curved arms at one and the same time, as one of the signals corresponds to the main line and the other to the branch line or siding.

In some forms of signal-detector the transverse rod **I** is joined on to a vertical bar which slides through guide-holes in a casting something similar to the arrangement shown in the casting **L**. Longitudinal guide-holes, parallel to the line of rails, are made in the casting a little above the transverse rod-bar, and through the longitudinal guide-holes slide two vertical bars which are attached to, and form part of, the wire connections to the two signals. The wire bars have each a small tongue or rectangular fin forged on to the under side of the bar, and there is one corresponding channel cut in the transverse rod-bar. When the switches are properly closed in one position, the channel cut of transverse bar will be opposite one of the wire bar fins, and will allow one of the signals to be pulled over, but the other wire bar cannot be moved. The closing of the switches in the reverse position moves the channel cut so as to allow the other wire bar to be pulled through, but as there is only one channel cut in the transverse bar, only one signal can be pulled over for each position of the switches.

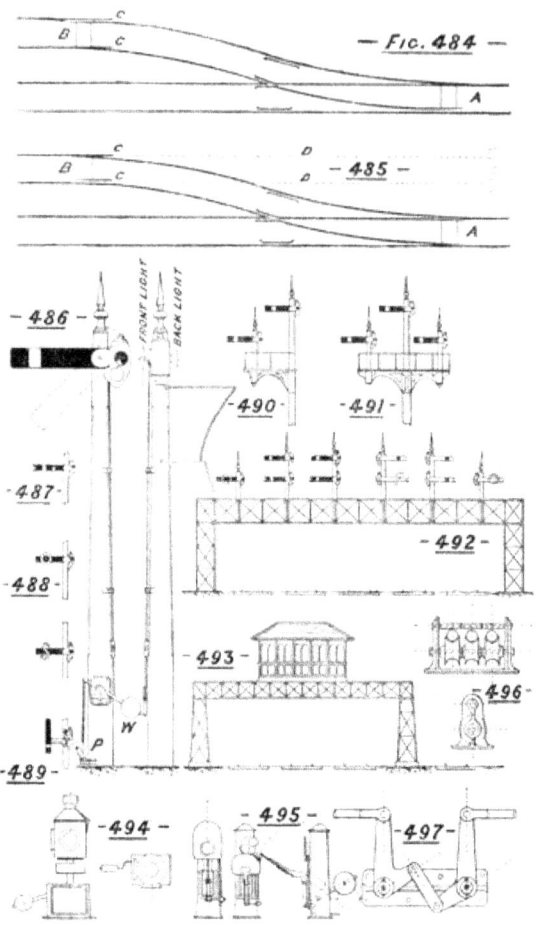

Throw-off or *trap points*, are introduced to throw an engine or train off the rails of a siding on to the ballast, and so avoid a collision with any other train which may be standing or passing on the line of rails with which such siding forms a connection. Fig. 484 is a diagram sketch of the arrangement, in which the main-line points are indicated by the letter **A**, and the trap points by the letter **B**; one series of rodding actuated by one lever in the signal-cabin works both the main-line points and the trap points at the same time and by the same movement. The connections are so made that when the points **A** are set for the passage of trains on the main line, the trap points **B** are set open to throw off on to the ballast, as shown in Fig. 484; and when the main-line points **A** are set to allow a train to pass from the siding on to the main line, the trap points **B** are closed, as shown on Fig. 485. A disc or other signal, worked or interlocked with the points, is placed near **B** to notify the engine-driver when he may pass out of the siding on to the main line; but should he from any cause proceed before the points are properly set and the corresponding signal given, his engine would run off at the ends of the rails **C, C,** and be derailed on to the ballast. The inconvenience caused by such derailment would be trifling compared with what might result from a collision with a train standing or passing on the main line. In some cases the siding is continued onwards for a considerable distance from the trap-point rails **C, C,** as indicated by the dotted lines **D D**, and terminates with a dead end. When this arrangement can be adopted, derailment is obviated, and the engine is brought to a stand by a buffer-stop at the end of siding. On no account should trap points be placed close to the top edge of a high embankment, or up to the abutment or wing walls of an under-line bridge, where an engine running through them accidentally might fall down a considerable height, and cause serious results. All sidings joining on to main lines should be trapped as above described, and when properly signalled and interlocked in the signal-cabin, the traffic-working can be carried on with increased facility and security.

Fig. 486 is a sketch of an average sample of an ordinary single-arm wooden signal-post, with signal-arm, lamp, spectacles, ladder, and gearing complete for wire connection to signal-cabin. When the arm stands out in the horizontal position, representing the *danger* or stop signal, the red spectacles will be in front of the lamps, and will show a red light to an approaching train. When the arm is lowered, as indicated by the dotted lines, the second spectacle will be in front of the lamps, and will show either a white or green light (according to the accepted code) as an *all-right* signal for the train to proceed. For many years a white light was adopted for the all-right signal, but latterly, to prevent confusion with other white lights about a station, there has been an increasing disposition to use a green light as an all-right signal. Several railway companies have already effected the change, and others have arranged to follow their example. The counter-weight **W** keeps the signal-arm to the danger position, except when it is raised by the pulling over of the signal-wire from the signal-cabin working over the pulley **P**. Should the wire break when being pulled, the weight **W** falls down to the stop-plate, and the signal-arm rises to danger. The signal-posts may be of wood, wrought-iron, steel lattice-work, or cast-iron.

The arms of *distant signals* should be cut to a fish-tail shape, as in Fig. 487, to distinguish them from other signals. Goods-line signals should have a thin sheet-iron ring, as in Fig. 488. Sometimes purple glass is used instead of red glass for the spectacles of goods signals. Letters or numbers may be attached to signal-arms to signify the lines or sidings to which they correspond. Special signals are sometimes made with the arm working on a centre pin, as in Fig. 489.

At junctions or places where two or three signals have to be fixed near together, it is customary to carry them on a bracket signal-post, as in Figs. 490 and 491. The former represents the home signals at an ordinary junction, the taller signal being for the main line and the lower one for the branch line. Fig. 491 shows the home signals at a junction where there is one line turning out of the main line to the left and another to the right. The taller signal in this case also serves for the main line and the two lower signals for the branch lines.

In important station-yards, where there are a large number of lines and sidings running side by side, it is not always convenient or possible to place the respective signal-posts in suitable positions between the lines. To overcome the difficulty, the signals are erected on light overhead lattice girders, as shown in Fig. 492. In some cases, for want of a better position, or to obtain a more comprehensive view of the lines and signals, the signal-cabin is built on lattice girders, as in Fig. 493.

Ground or *disc signals* are fixed at the ground-level, and are worked in conjunction with trap points or outlet switches from sidings. In some cases they are worked direct by a connecting-rod from the switches, and serve merely as indicators to show whether the switches are lying for or against an engine passing out of the siding. In other cases they are worked independently from the signal-cabin by a separate lever and wire connection, the interlocking being so arranged that the lever working the switches must be pulled over before the lever working the disc signal can be moved. In one type the disc signal is fixed to a short vertical axis, as shown in Fig. 494, and by means of a cranked arm is made to rotate a quarter of a circle, so as to exhibit either a stop or advance signal according to the position in which the switches are lying. In another type, the lamp is fixed, and the red disc, with a red glass in the centre, is made to assume a horizontal or vertical position by a rod and crank, as shown in Fig. 495.

A simple arrangement of rodding and rollers for switch connections is shown in Fig. 496, the number of sets of rodding being determined by the number of connections to be made. Fig. 497 is a rodding compensator, to compensate or adjust for the difference in length of the rodding arising from variations in the temperature. The compensator may be used either vertically or horizontally, according to space or circumstances.

Strong wrought-iron or steel cranks of different angles will be required when changing the direction of the rodding, or connecting to switches and facing point-locks. They must be firmly secured to strong timber framework well bedded in the ballast. For cranks working switches and bolt-locks, it is better to use extra long timbers under the rails instead of the ordinary sleepers. Cross-pieces can be bolted to the ends of the long timbers, and the cranks placed practically on the same timbers carrying the permanent way. By this means the rails and cranks can always be maintained in their proper relative positions as to distance, line, and level.

Without a large series of diagrams it would be impossible to adequately describe the extent of signalling and interlocking required at large terminal stations and important roadside stations, but one or two simple examples may serve to illustrate the general principles.

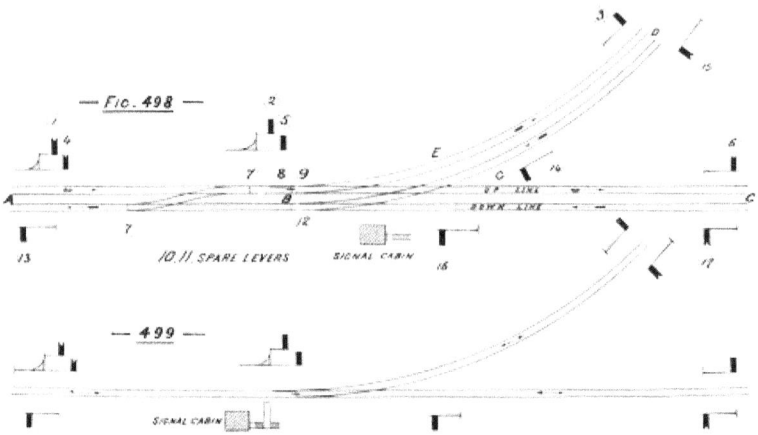

Fig. 498 represents the modern grouping of signals considered necessary at an ordinary double-line junction, showing all the signals at their normal or *danger* position. The numbers marked on each indicate the numbers of the levers in the interlocking frame of the signal-cabin. Four distinct sets of trains have to be dealt with at this class of junction, and the interlocking must be so arranged that when the signals are lowered for the advance of any one train, no conflicting signals can be given to any other train.

Assuming a train approaching from **A**, which has to continue on the main line past **B** on towards **C**, then the levers in the signal-cabin must first be pulled over to set the points 9 and bolt-lock 8 in proper position for the main line; and this operation will release the levers which have to be pulled over to lower the signals 5, 4, and 6, but at the same time will lock, and prevent the pulling over of the levers or lowering of the signals 2 and 1 for a train from **A** to **B** and **D**, or of the signals 14 and 15 for a train from **D** to **B** and **A**. The levers will, however, be free to pull over for setting the points 12 and lowering the signals 16, 17, and 13 for a train on the main line from **C** to **B** and **A**.

In a similar manner, assuming a train approaching from **D**, which has to continue up to the main line at **B** and on towards **A**, then the lever in the cabin must first be pulled over to set the trailing points 12 in proper position; and this operation will release the levers which have to be pulled over to lower the signals 14 and 15, but at the same time will lock, and prevent the pulling over of the levers or lowering of the signals 5 and 4 for a train from **A** to **B** and **C**, or of the signals 16 and 17 for a train from **C** to **B** and **A**. The levers will, however, be free to pull over for setting the points 9 and bolt-lock 8 and lowering the signals 2 and 1 for the passage of a train on to the branch line from **A** to **B** and **D**.

For a train from **C** to **B** and **A**, the levers 12, 16, 17, and 13 would be required, and these would lock levers 14 and 15, and prevent the approach of any train from **D** to **B**, but they would leave free the levers necessary either for a train from **A** to **B** and **C**, or for a train from **A** to **B** and **D**, but only one of them at a time, the setting of the one series locking the other series.

A train from **A** to **B** and **D** would require the proper setting of the points 9, bolt-lock 8, and signals 2, 1, and 3; and these would lock 5 and 4, but would leave free the levers necessary either for a train from **C** to **B** and **A**, or for a train from **D** to **B** and **A**, but only one of them at a time.

The cross-over road from the UP to DOWN main line, near the letter **B** on sketch, is only intended for use in case of break-down or accidents, and the normal position of the points is to lie clear for the passage of trains on the main lines. To use the cross-over road, the whole of the signals must first be set to *danger* before the points 7 and 7 can be opened to permit the passage of an engine or train from the one main line to the other.

The starting signals 6 and 3 should be placed sufficiently far away that the longest passenger or goods train may stand between them and the clearance points at **G** and **E**. These starting signals are of great service to train-working at junctions. Supposing a main-line train from **A** arriving at **B** before the section from **B** to **C** was clear, such train could be brought to a stand at signal 6, and remain there while another train from **A** was allowed to pass **B**, and proceed onwards towards **D**; or a branch-line train from **A** to **D** could be brought to a stand at 3, to allow a main-line train to proceed onwards from **A** to **B** and **C**. The starting signal 13 should be placed well in advance of the cross-over road to control anything passing from one line to the other.

Fig. 499 shows the modern grouping of signals for an ordinary single-line junction. The arrangement is almost practically the same as for the double-line junction shown in Fig. 498, there being the same four distinct sets of trains to be controlled, but not any cross-over road. The signal-cabin is placed on the main line, a little in advance of the facing points, and a well-fenced-in gangway, the same height as the engine footplate, is carried out the proper distance from the rails, on which the signalman can stand to hand over or receive the train staff from the engine-driver when passing.

At stations and places where there are several sidings and lines connecting with the main lines, at considerable distances apart, it will be necessary to have two or more signal-cabins placed in suitable positions, not only for expediting the working of the constant shunting movements, but also to insure that there is a signal-cabin within the regulation distance of all facing points on the main line. So far as the main line is concerned, the interlocking of these cabins must be connected, the one with the other, by slotting, or co-acting gearing, in such manner that the cabin in advance shall always be able to control the cabin in the rear in the lowering of the main-line signals for an approaching train. Fig. 376 is a diagram sketch of a typical double-line roadside station with two signal-cabins. The NORTH cabin has to work the signals and points in connection with the goods-shed, goods-sidings, and market branch, and the SOUTH cabin, those in connection with the coal and cattle sidings; and each of the cabins to work the signals and points of that portion of the main line adjoining its own cabin. For siding working, each cabin is quite distinct and independent of the other, but for main-line working the lowering of the signals can only be effected by the joint operation or co-acting of both cabins.

Assuming a train approaching from **A** to proceed in the direction towards **B**, then, before the signalman in the NORTH cabin can lower the UP home-signal **C**, the signalman in the SOUTH cabin must first pull over his lever and release the slot which retains the signal **C** at *danger*, and in doing so the levers in his own cabin will stand locked, and prevent the lowering of the signal **D**, or opening of points **E** to allow access from the sidings to UP main line. The cross-over road **F G** will also be locked for main line clear. When the slot has been released from signal **C**, the signalman in NORTH cabin can lower the UP home signal **C**, but before he can pull over the lever for this purpose he must first lock the points **H**, to prevent access from the sidings to the UP main line, and also the points **K L** of the cross-over road, to keep the main line clear. A similar operation has to be gone through for a train approaching from **B** to proceed in the direction towards **A**, when the signalman in NORTH cabin must first withdraw the slot from the DOWN home-signal **M** before the signalman in the SOUTH cabin can lower that signal. A small automatic disc is placed in the cabin to indicate to the signalman when the slot has been withdrawn by his colleague in the neighbouring cabin, and for facility of working, the two cabins are usually placed in communication with each other by telegraph or telephone.

At some stations similar to the above, where there is a very frequent train service, with several of the trains running through without stopping, it is the practice to have a second or lower arm to the home signals **C** and **M**, as shown on the diagram, these lower arms being only *pulled off* for through or non-stopping trains, as an indication to the engine-driver that the line is clear in the section ahead.

In addition to the leading signals shown in the sketches, there are shunting signals for the movement and marshalling of trains—setting-back signals in connection with the making up of passenger trains; taking on or off passenger carriages; or moving out empty passenger carriages; and many other special signals which become necessary for the working of a large and complicated train service.

The above simple diagrams will explain some of the principal requirements to be kept in view when working out signalling arrangements. Where the lines and sidings are very numerous, as at important junctions and large terminal stations, the signalling becomes very intricate, and may require three or four cabins, slotted together in such manner that the necessary co-acting may be insured for the proper controlling of the mainline signals. Many of these signal-cabins contain a large number of levers, some of them having as many as a hundred, and a few of them two hundred and forty levers, or more, all of them so carefully arranged that no conflicting signal can be given. Not only has much skill to be exercised in the accurate adjustment of the interlocking machinery, but much study must be devoted to determine the exact duty of every lever, for the locking or releasing of other levers.

Signal-cabins may be built of stone, brick, or wood. They should be roomy, well ventilated, and have abundance of light. Every signal-cabin should be placed in the position from which the signalman can obtain the best view of the signals and points under his charge. The height of the cabin floor will depend upon any obstacles that may intervene between the cabin and the signals, such as over-line bridges, station roofs, buildings, or other obstructions. Sometimes the floor has to be kept as low as five feet above rail-level, to secure a line of sight under the over-line bridges; and in others the floor has to be raised twenty, or even thirty feet above rail-level.

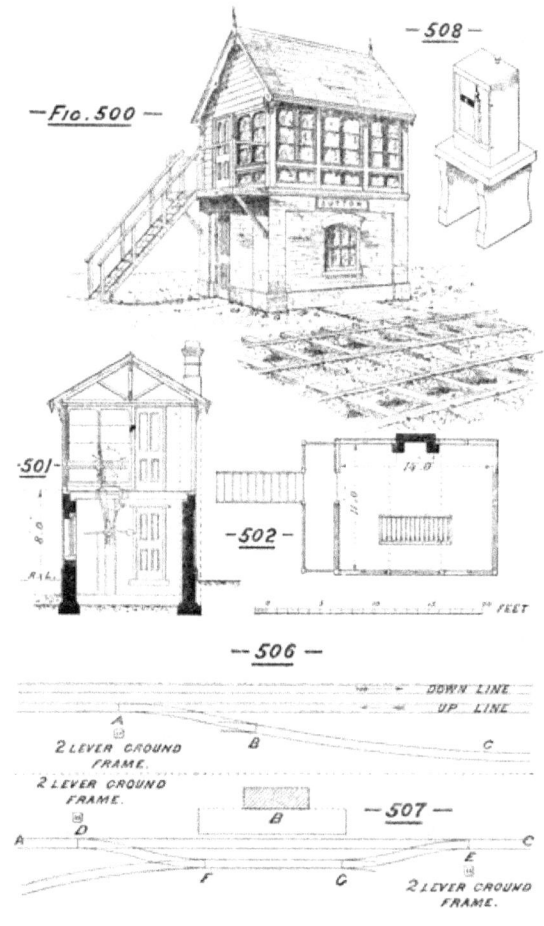

Figs. 500, 501, and 502 show plan, transverse section, and elevation of a signal-cabin suitable for a small roadside station. The lower story and chimney-stack are of brick, and the upper story of wood, with slated roof. There is room for an interlocking frame of twenty or twenty-five levers, and space at the end of the cabin for the block-telegraph instruments, or electric train-staff instruments. The roof-work is open up to the slateboards, to obtain as much air capacity as possible. In the transverse section a winch for working mechanical gates is shown at the end of the interlocking frame. There is a liberal amount of glass, and two or three sliding windows, which the signalman can open to enable him to speak to the engine-drivers or others during shunting operations. The lower story of the cabin can be utilized for trimming lamps and keeping a small supply of coals and other stores. When working after dark the lamps in the cabins should be well protected by shades, to prevent the lights being seen by engine-drivers, and mistaken for signals.

Interlocking.—There are several systems of interlocking, each of them varying considerably in the form and mode of application, but all of them having the same general object of securing or releasing the necessary levers for each combination of signalling movements. A brief description of one of the systems will explain the order in which the movements have to be made, and the security which can be obtained by the locking.

Figs. 503, 504, and 505, are sketches illustrating one of the types of wedge and tappet interlocking. Each lever works on a fulcrum or pinion as at **A**, and has a lower arm **B** for lifting the rods leading off to points or signals, and an arm **C** to carry a counterweight when necessary. Cast-iron braces **D** are placed at convenient distances between the series of levers to carry the top frame **E** on which the lever floor casing **F** is bolted. This casing is continuous from end to end of the locking frame, with the exception of the narrow openings through which the levers travel when moving backwards or forwards. The sleeve-block **G**, resting in the depressed portions of the arc, retains the lever in position. When taking hold of the main lever **L**, the signalman's hand draws the small side lever **M**, close to the main lever, and raises the sleeve-block **G** sufficiently high to pass over the top of arc **F**, the lever **L** can then be pulled or pushed over, and the block **G** will fall into the depression at the end of the stroke when the hand is removed. **N** is a tappet or thin flat bar attached to the main lever, and which works backwards or forwards between the wedges in the wedge frame **O**. The wedges move horizontally between guide pieces, and work either singly or are connected by the lower slide bars to other wedges some distance away on the frame according to the position of the levers which have to stand or move in unison for the releasing or locking. A strong cover is placed over the wedge frame to keep out the dirt.

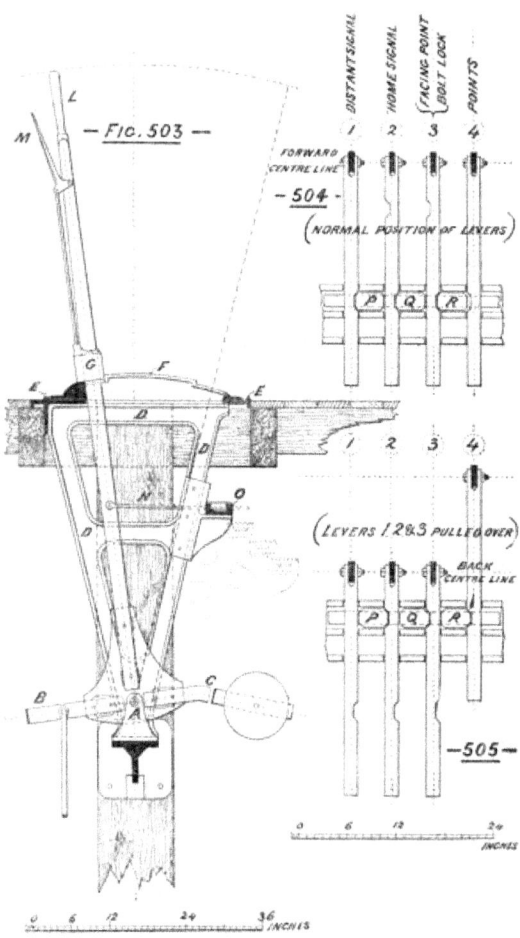

Figs. 504 and 505 show plan views of four levers in a signal cabin taken just above the level of the tappets. In Fig. 504, all the levers are in their *normal* or forward position, with the home and distant signals at *danger*, and the facing points leading into loop or siding lying for main line. Previous to the approach of a train on the main line, the home and distant signals have to be lowered, and will require the pulling over of levers 1 and 2; but these levers cannot of themselves be moved, as the wedges **P** and **Q** are locked by the straight side of lever 3. The operation would therefore be as follows:—points lever 4 being set in its normal position for the main line would remain forward, lever 3 working the facing point bolt-lock would be pulled over, and in doing so would move the wedge **R** to the right into the recess of tappet of lever 4, locking that lever, and presenting the recess of its own tappet ready to receive the wedge **Q**. Lever 2 can then be pulled over, and will move the wedge **Q** to the right into the recess of tappet of lever 3, and present its own recess for wedge **P**. The pulling over of lever 1 completes the series, by moving the wedge **P** over to the right into the recess of tappet of lever 2. Fig. 505 shows the positions of the tappets and wedges with the levers 1, 2, and 3, pulled over to make the combination described. Upon examination, it will be seen that levers 2, 3, and 4, are all securely locked, the points cannot be moved, nor the facing point bolt-lock withdrawn, nor the home signal changed until the lever 1 is pushed over again into its normal or *danger* position. To restore the levers to their forward position, they must be set back in the reverse order to which they were pulled over. To simplify the explanation, only four levers are shown in the above sketches, but the principle is constantly extended out to a very large number of levers, and in many cases necessitates the introduction of several rows of wedges as indicated by the dotted lines. In some instances a combination is effected by pulling a certain lever only half over. In some systems the preliminary action or spring handle locking is adopted, in which the locking is actuated by the small side lever, similar to the one marked **M** on Fig. 503. The advocates of this arrangement claim increased security and precision in the interlocking, while on the other hand it is alleged that the mechanism is rendered more complicated without any corresponding advantage.

Detached Lock.—Sometimes there is in the vicinity of a railway station, a siding which is too far away to be worked direct from a signal cabin, and not sufficiently used to warrant a separate cabin. Such sidings can be worked by a small ground frame opened or locked by a special key attached to the interlocking machinery in the adjoining signal cabin on a double-line railway, or attached to the train staff on a single line.

Fig. 506 shows the arrangement applied to a double line with the outlying siding turning out of the UP main line, the points lying in a trailing direction for the running trains. Before the special and *only* key can be withdrawn from its seat in the interlocking frame of the signal cabin, all the UP main line signals must be set to *danger*, and cannot be moved from *danger* until the key is restored to its proper seat again. When the key is removed from the signal cabin, it can be taken to the ground frame at **A**, inserted in the key opening, and by turning it partly round, will release the bar which locks the levers of the facing point bolt-lock and the points. When these two levers are free the points can be opened, and vehicles moved into or out of the siding **B C**, but the special key cannot be withdrawn from the ground-frame **A**, until the points and facing point bolt-lock are put back again into their normal position for main line working. When the operations at the siding are completed, the special key can be removed, and taken back to its proper place in the signal-cabin, and ordinary working be resumed.

Fig. 507 shows the application of the detached locks on a single line, and is a sketch of a portion of railway on which there is a small station **B**, with a goods siding **F G**, where the traffic is too small to require anything more than ground frames and detached locks. An engine-driver before leaving the station **A**, receives a train staff, which gives him possession of the line as far as **C**, including of course the intermediate station **B**, and this staff he must carry with him and hand over to the signalman on his arrival at the end of the section at **C**. At each of the points **D** and **E** is placed a two lever ground frame, similar to the one shown in Fig. 506, and attached to the train staff is a key, which will operate either of the two ground frames, but only one at a time, as the key must be inserted before the levers can be moved. When the train is proceeding in the direction from **A** to **C**, it will be more convenient to shunt vehicles into or out of the siding **F G**, by means of the points **E**, but when proceeding from **C** to **A**, the points **D** will be more convenient. Whichever of the points be used, they must be set, and bolt locked for the main line before the train staff and its key can be withdrawn from the ground frame and restored to the engine-driver. As the siding is *trapped* at **F** and **G**, it is impossible for any vehicles to be moved out on to the main line except through the medium of the train staff and key. The same arrangement of detached lock is equally available for a single siding with only one set of points.

Electric Repeater.—It will sometimes occur that on account of a curve or other obstacle, the arms and back lights of a distant or other signal cannot be seen from the signal cabin, and it is necessary to introduce an electric repeater. This little instrument consists of a miniature semaphore signal fixed in a metallic box with a glass front, and placed on a stand about a foot above the floor level immediately in front of the signal lever for which it is intended to serve as an indicator. Like the signal proper, the normal position of the miniature semaphore is at *danger*, but when the signal lever is pulled over in the cabin, the rod that pulls down the arm on the signal post effects a contact with an electric circuit which lowers the arm of the miniature semaphore at the same moment that the signal arm proper is lowered, and gives visible indication in the cabin that the signal is working. Fig. 508 is a sketch of one form of electric repeater.

Detonators or fog signals are largely used in foggy weather and snowstorms, when the out-door signals cannot be seen from an approaching train. At such times the atmosphere is so dense, and the surrounding objects so obscured, that an engine-driver is totally unable to distinguish the usual landmarks which guide him on the approach to a station or semaphore, and he might easily pass by a signal unless he received an audible signal to indicate the position of the one that is invisible. Detonators are usually made in the form of a circular tin or metallic case about two inches in diameter, and three eighths of an inch thick, with soft metal clips on opposite sides for bending over and securing to the rails. The case is filled with detonating powder, which is crushed by the first wheel passing over it, and explodes with a loud report. It is customary to use these detonators in pairs placed a short distance apart in case one of them should fail to explode.

Fog-signalling regulations vary on different railways, but the arrangements are generally carried out somewhat in the following manner. During the prevalence of a fog or snowstorm, a fog signalman is placed near each of the signal-posts to be protected, and is supplied with a hand signal-lamp, hand-flags, and a packet of detonators. So long as the arm of the signal-post at which he is alongside stands at *danger*, he must keep two detonators on the rail of that line which the signal controls, and also show a RED hand-signal (hand-flag by day, and hand-lamp after dark) to the approaching train. When the signal arm is lowered to show that the line is clear for the passage of the train, the fog signalman must remove the two detonators, and show a GREEN hand-signal (flag, or lamp) to the approaching train. When an engine driver hears the report of a detonator crushed by his engine, it is his duty to shut off steam immediately, and bring his engine to a stand, after which he must proceed very cautiously, until he receives further signals by hand or otherwise, or receives the line-clear signal to continue on his journey. Detonators are also of great service both in fine or bad weather, in cases of a wash away, a failure of works, or obstruction on the line, when a hand-signal may not be seen, but a detonator must be heard.

Mechanical Gates.—Mechanical gates, worked and controlled from the inside of a signal-cabin, are now very largely adopted for public road level-crossings instead of ordinary hand-gates, opened and closed by a gateman walking from side to side of the line across the rails. Being worked from inside the cabin, they remove all possibility of the gateman being struck by a passing train; they move simultaneously, and can be opened or closed in very much less time than hand-worked gates, which have to be moved one by one, and being interlocked with the signals, the mechanical gates cannot be placed across the lines of rails until the train-signals in each direction are set at *danger*. When set for either train traffic or public road traffic, the gates are held firmly in position by metal stops, rising out of cast-iron boxes lying flush with the ground, and worked by a separate lever in the signal-cabin.

Assuming the gates to be set for train traffic, and it is desired to open them for the public road traffic, the first operation will be to pull over the levers, and raise the signals in each direction to *danger*, and thus release the stop-lever, which can then be pulled over, to lower the gate-stops and allow the gate-winch to be turned, and the gates moved round into correct position. The stop-lever must then be set back to raise the stops and hold the gates secure. The train-signals will be retained at *danger* by the interlocking gearing, and cannot be lowered until the gates are set back again across the public road, and the gate-stops raised.

It is frequently urged that the celerity with which mechanical gates can be swung round and closed across the public road, is in itself a source of danger, and that persons preparing to cross the line might be struck by a moving gate, unless they received a distinct warning that such closing was about to take place. There is no doubt persons have been struck by such gates when closing across the road, and heavy claims for injuries have been decreed against railway companies, who were unable to prove that the man in charge had called out or given warning before moving the gates. To ensure that due and undeniable warning shall always be given, a firm of signal-makers have patented an appliance by which a powerful electric gong, fixed on the top of a tall post close to the gates, is sounded automatically by the gate machinery itself, and before the gates actually commence to move. As previously described, the pulling over of the lever to lower the gate-stops is the first operation to be performed whenever it is necessary to change the position of the gates, and it is the pulling over of this lever which actuates the apparatus, by bringing two electric points into contact, and thus starting the ringing of the gong or alarm. The gong continues to sound until the gates are moved over, the gate-stops raised, and the stop-lever put forward again into its normal position. The arrangement is very simple and very effective, and being purely automatic must work as regularly as the stop-lever. The tone and volume of the gong can be varied to suit circumstances. The public soon become familiar with its sound, and recognize its meaning.

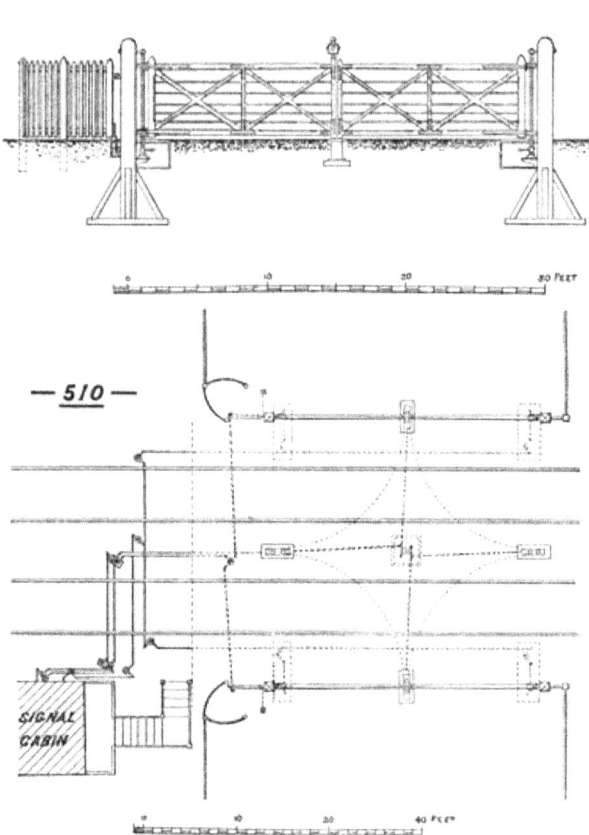

— FIG. 509 —

— 510 —

Figs. 509 and 510 give sketch plan and elevation of a set of mechanical gates for a public road level crossing on a double line of railway. The signal-cabin should be placed within a few yards of the gates, to enable the man in charge to have a good view of the persons and vehicles passing over the roadway. The underground gearing for working the gates and stops, must be protected by iron or wooden casing. The swinging portion of the wicket gates is closed, and held by a separate lever. The gates shown on the sketch are for a crossing on the square, but they can be equally well arranged for an oblique crossing, and of widths to suit the locality.

Block-Telegraph Signalling.—However complete the outdoor signals and interlocking at any station, they can only control the movement of trains within their range, and something more is requisite to ensure the safe working of the traffic over the long lengths of line between stations. For some years a time-interval was allowed for the working of trains following one another on the UP and DOWN lines of a double line railway, no train being allowed to leave a station sooner than a fixed number of minutes after a previous train had started in the same direction. With this system there was always the risk that the first train might be overtaken and ran into by the second, and especially in the night time, or when the atmosphere was at all foggy. The electric telegraph was then called in to assist in the train-working, and brief telegrams were passed between the stations announcing the departure and arrival of trains. The increased security and convenience thus obtained led to the introduction of special electric telegraph instruments, devoted to the exclusive duty of train-working. These instruments, termed block telegraph instruments, are now almost universally used on all double lines of railway, and have largely contributed to the safe and efficient working of an ever increasing traffic. They are made in various forms, but the object of each is to ensure that before any train is allowed to start from, or pass any station, the signalman at that station shall receive from the signalman in the cabin in advance a distinct visible signal that the line is clear, and free of any train up to the cabin in advance; and also that after the train has been despatched, the signalman in the rear shall be at once advised when the train has arrived at the signal-cabin in advance. Fig. 511 is a sketch of one type of block-telegraph instrument, in which the leading feature is the miniature signal-post with its two arms, an arrangement which readily appeals to the eye of the signalman as being so similar in form and action to the fixed signals in the station. Each instrument is supplied with a bell or gong, by which the adjacent signalmen can communicate with each other, in accordance with a fixed code of signals which defines the relative numbers of strokes of the bell or gong, to represent certain regulation calls and answers. In the signal-cabins of the intermediate stations, two block-telegraph instruments are required, one for the section of the line to the left hand of the cabin, and the other for the section to the right. At the terminal stations only one instrument is required.

In the instrument shown in Fig. 511, the upper arm of the miniature signal-post is coloured RED, and is moved by electricity through the medium of the block telegraph instrument in the signal-cabin in advance; and until this RED signal be lowered to the *line clear* position by the signalman in the cabin in advance, no train must be allowed to start from or pass the cabin in the rear. The lower signal-arm coloured WHITE is lowered by the plunger **A** on its own instrument by the signalman in charge, and at the same moment lowers by electricity the upper or RED arm of the block-telegraph instrument in the signal-cabin at the other end of the section. The lower or WHITE arm is thus restricted to the signals sent away from the signal-cabin, while the upper or RED arm is restricted to signals received in the signal-cabin. In the centre there is a round handle **B**, which rotates a circular disc inside the instrument, and on this disc are painted three distinct train inscriptions, only one of which can be seen at a time through the glazed opening. One inscription has the words ALL CLEAR painted in black letters on a WHITE ground; another has the words TRAIN ON LINE painted in white letters on a RED ground; and the third has the words TRAIN OFF, BUT SECTION BLOCKED painted in black letters on a GREEN ground. The instrument is considered to be in its *normal* position when the GREEN inscription is in view, and both the miniature signal-arms raised to *danger*.

Fig. 512 represents a portion of double line divided out into sections, or working blocks, between the stations **B**, **C**, and **D**. Each station is provided with the necessary block-telegraph instruments, and the usual distant, home, and starting semaphore signals.

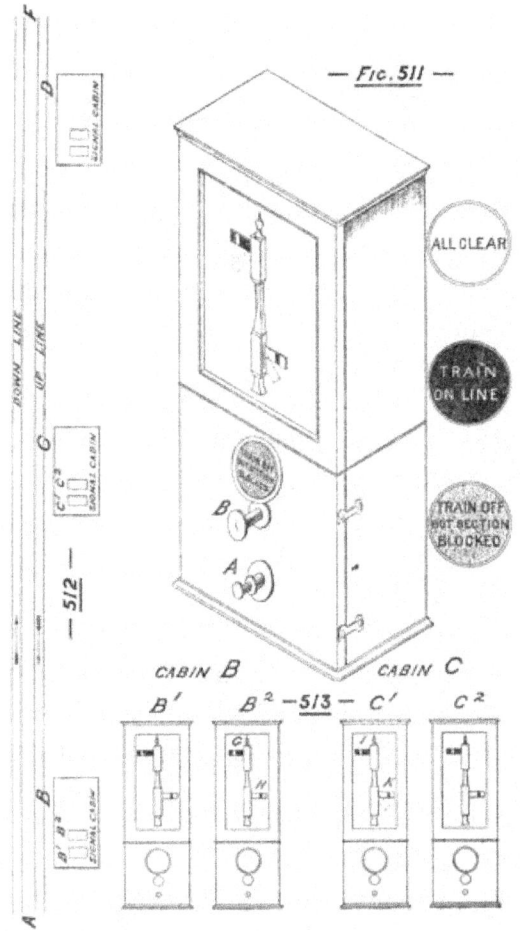

Fig. 513 is a diagram sketch showing the pair of instruments as they stand on the instrument-tables in the signal-cabins **B** and **C**, where B^2 and C^1 are the instruments which work together for the block section **BC**. Supposing a DOWN train proceeding from **A** in the direction of **F**, and approaching the signal-cabin of the block station at **B**, the DOWN starting signal standing at *danger*; then by the code of signals on the bell or gong the signalman at cabin **B** would communicate with the signalman at cabin **C**, to obtain *line clear*, so as to allow the approaching train to proceed on to **C**. If the previous train in the same direction had already passed **C**, and there was not any obstruction on the line, the signalman at **C** would give *line clear* for the DOWN train, and to do so he would turn his circular disc to show the WHITE inscription ALL CLEAR, and then push in the plunger of his C^1 instrument, lowering the DOWN or white arm, **K**, of his own instrument to the position shown by the dotted lines, which operation would at the same moment lower by electricity the DOWN or red arm, **G**, of the instrument B^2 in cabin **B** to the position of the dotted lines. The signalman at **B** would then lower his starting signal, to allow the DOWN train to proceed on towards **C**, and immediately the train had passed the starting signal he would, by means of his bell or gong advise the signalman at **C** that the train had entered the section, or block **BC**, and the signalman at **C** would at once turn his circular disc to show the RED inscription TRAIN ON LINE, and use his plunger to raise to *danger* the DOWN or white arm, **K**, of his own instrument, and at the same time raise by electricity the DOWN or red arm, **G**, to danger in the instrument B^2 in cabin **B**. The section **BC** would then remain blocked until the DOWN train had arrived, or passed the station **C**, when the signalman there would, by means of his bell or gong advise the signalman at **B** that the DOWN train had passed out of the section, and would turn his circular disc to show the GREEN inscription TRAIN OFF, BUT SECTION BLOCKED. Both instruments would then be in their *normal* positions, with the arms raised to danger, and ready for further train operations. In a similar manner for the UP-line trains on the section or block between **C** and **B**, the signalman in **B** cabin would turn his circular disc, and use his plunger to lower the UP or white arm, **H**, in his own instrument, B^2, and at the same moment lower by electricity the UP or red arm, **I**, of the instrument C^1 in cabin **C**, the other operation for train on line and train off being carried out for the UP train in the same routine as for the DOWN train. The outdoor fixed signals, or distant home and starting semaphore signals, have all to be worked to correspond to the block telegraph signals, and as the latter are always received well in advance of an approaching train, it follows that when the line is clear, the outdoor signals can be lowered so as to allow a through or non-stopping train to pass a block-telegraph station at full speed.

Where the traffic is moderate, it may be sufficient to have block-telegraph instruments at each of the stations, but with a very frequent train service it will be found necessary to divide the line into shorter sections, and erect signal-cabins and block-telegraph instruments at intermediate points between stations.

The code of bell or gong signals is extended to include various matters in connection with the train-working. For example, when a DOWN train is passing cabin **B** at full speed, the signalman may observe that there is something wrong—a carriage or waggon on fire, a tail-lamp missing, or other irregularity. It is too late to stop the train with his own signals, but by means of his bell or gong he can call upon the signalman in cabin **C** to stop and examine the train, and the DOWN distant and home signals at **C** can be raised to *danger* before the train reaches the cabin at **C**.

In every block-telegraph signal-cabin there is a train-book in which the signalman has to write down the time and description of every arriving or passing train, and, as this book lies before him, he has a complete record of the train-working, with the particulars of the exact times when the *line clear* signals were given, and also when the train arrived or passed his signal-cabin.

To guard against the possibility of a signalman inadvertently giving *line clear*, or allowing another train to pass his cabin before the previous train had reached the signal-cabin in advance, some railways have adopted the lock and block system. By this arrangement the starting signal at any cabin is electrically and mechanically locked from the cabin in advance, and can only be released or lowered by the action of the outgoing train itself when passing over a treadle or other appliance connected with the rails of the running-line at the signal-cabin in advance. This method practically gives the train the complete control of the section; and any signalman attempting, in error, to lower his starting signal would find it to remain fixed to *danger* and immovable, until released by the arrival of the train at the advance cabin.

Train-staff for Single Line.—When there is only a single line of railway for both an UP and DOWN train-service, very definite precautions must be adopted to prevent the meeting or collision of trains travelling in opposite directions. Where the piece of single line is short, and can be worked by one engine in steam, or two coupled together, no collision can take place, as the train-service will be limited to the one train moving backwards and forwards over the section; but with a long length of single line, including a large number of stations, necessitating several trains, some clear and comprehensive regulations must be introduced. For a long time the simple train-staff was found to give the desired security; there was only one staff for each pair of adjoining staff-stations, and no train was authorized to run without the staff, and as the staff could only be on one train at a time, the precaution against collisions was looked upon as complete. These staffs, which were generally made of brass, or other metal, were sufficiently large to be conspicuous when placed in the stand prepared for them on the engine. They were lettered to correspond to the stations to which they belonged, and were made in different patterns to distinguish them for their respective sections. No train was allowed to start from a station until the engine-driver received from the station-master the proper staff to authorize him to proceed to the next station, and on his arrival there it was the duty of the engine-driver to hand over the train-staff to the stationmaster of that place, and wait for another train-staff to authorize him to proceed over the next section. So long as the train service could be evenly arranged, and that there was always an UP train to take back a train-staff which has been carried out by a DOWN train, the simple staff worked most efficiently; but as the traffic increased, and two or more trains had to be despatched in the DOWN direction before one had to run in the UP direction, some auxiliary arrangement had to be introduced. This was effected by issuing train tickets, kept in a locked-up box, which could only be opened by the key attached to the train-staff. A properly dated train-ticket was handed to the engine-driver of the first DOWN train, and, if necessary, a second train-ticket to the engine-driver of a second DOWN train, and then the train-staff itself was handed to the engine-driver of the third DOWN train. There were one or two serious drawbacks to this train-staff and ticket-working. As there was only a time interval between the starting of the trains, the one train might overtake and run into the other with disastrous results. Again, a second or third train, which was put down in the schedule, might be withdrawn at the last moment, and the staff left behind at a station when it was required at the opposite end of the section, thus causing much confusion and delay. The ordinary electric telegraph could have been utilized to assist in regulating these train movements, but it was felt that a mere telegraph message was not sufficient to ensure positive safety, and that something more tangible was required in the shape of a staff, or token, without which no train should be allowed to travel on a single line of railway. To meet this requirement, the electric train-tablet, and the electric train-staff instruments have been invented, each of them being so arranged that upon any one section, or pair of instruments, a tablet or train-staff may be taken out from the instrument at either end of the section, but when once taken out, no other tablet or train-staff can be withdrawn from either instrument until the first has been delivered and placed again in one or other of the two instruments.

Figs. 514, 515, and 516 are sketches of an electric train-staff instrument which has been very largely adopted on single lines, both at home and abroad.

In a similar manner to the block-telegraph instruments for double line, the electric train-staff instruments have each a bell or gong by which the adjacent signalmen can communicate their calls and answers in accordance with a regulation code. In the signal-cabins of the intermediate stations two instruments are required, one for the staffs belonging to the section to the left of the cabin, and the other for the staffs of the section to the right. At a terminal station only one instrument is required.

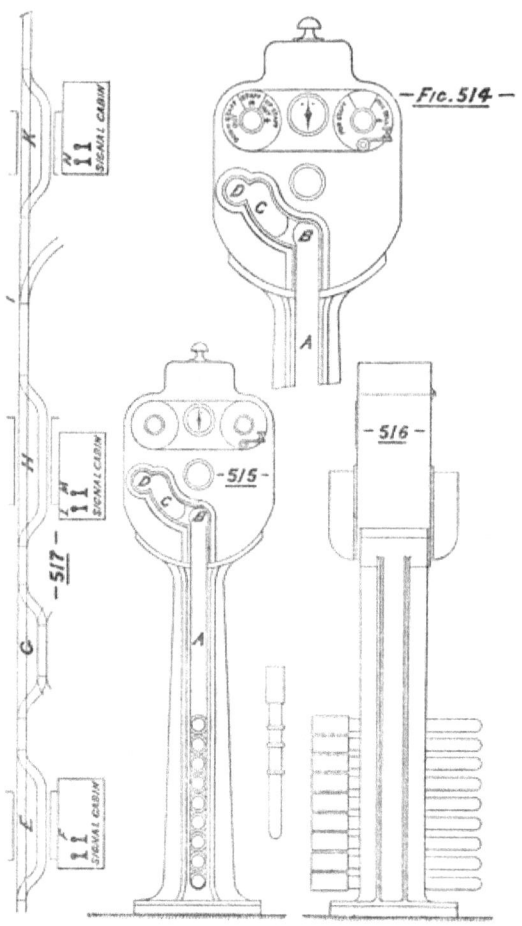

The head of the instrument contains the electrical and mechanical locking apparatus which controls the withdrawal of a train-staff, or is acted upon by its insertion. The circular name-plates and pointers, together with the galvanometer in the centre, serve as indicators to guide the signalmen in carrying out the various operations. The staffs usually consist of thin steel tubes, solid at the ends, with metal rings fixed upon them, as shown in the sketch, the number and position of the rings varying according to the section or pair of staff stations to which they belong; this difference in the rings effectually preventing the possibility of one set of staffs being used or inserted in either of the instruments of the adjoining sections. The staffs rest normally in the long vertical slot **A**, with the rings fitting in vertical grooves,

358

which prevent the removal of any staff except by passing it along the curved slot **BC**, and out by the opening **D**, of large diameter. The electrical and mechanical locking apparatus is placed at the curved slot, and until the locking-bolt, which stands across the passage of the curved slot, be lifted by the joint operations of the signalmen and their instruments at both ends of the section, no staff can be withdrawn. When the instruments are standing in their *normal* position of "staffs in," the signalmen can arrange between them to withdraw a staff—say either from the NORTH cabin instrument or from the SOUTH cabin instrument of the section, but only from one of them; and the act of taking out that staff automatically locks both instruments, and prevents the possibility of taking out any other staff from either instrument until the staff already removed is restored and inserted in one or other of the instruments. From the above description it will be seen that the electric train-staff instrument provides for the safe working of two or more trains proceeding, one at a time, in the same direction over a section of single line, each one being supplied with a train-staff, which must be handed over at the end of a section before another staff can be issued for a following train. Should the train-staffs accumulate in one instrument, in consequence of more trains running in one direction than another, a re-distribution of staffs is effected by the authorized persons according to fixed regulations.

In the diagram sketch, Fig. 517, a piece of single line is shown divided into sections or blocks, with loops or passing-places at the stations. At the station **E** a train-staff taken out of the instrument **F** serves for the section up to the instrument **L** at the station **H**; and on the train-staff is a key which will open the detached locks on the points of the small intermediate station, **G**, as described in Fig. 507, in connection with the working of detached locks. At the station **H** the engine-driver receives another staff from the instrument **M**, which takes him to the instrument **N** at station **K**, and in like manner on this staff is a key which will open the detached lock on the colliery siding points at **I**. At stations **H** and **K** are shown loops, or short pieces of double line, with platform to enable an UP train to cross or pass a DOWN train. The distance apart of the electric train-staff stations will depend greatly upon the number of the trains, and for a frequent train-service it may be necessary to have the instruments at every station, whether large or small. The electric train-staff is of great advantage in the working of ballast or construction trains, as a staff may be taken out of the instrument **F** at station **E**, which will give possession of the section as far as station **H**, and when the ballasting operations—which may be very near to **E**—are completed, the train can return to **E**, and deliver the staff again to the instrument **F**, instead of having to run the entire distance to station **H**. Although carrying a train-staff, the engine-driver must approach stations cautiously, and obey the fixed signals in the usual manner.

CHAPTER VIII.

Railways of different ranks—Progressive improvements—Growing tendency for increased speeds, with corresponding increase in weight of permanent way and rolling-stock—Electricity as a motive-power.

Looking at railways in their present stage of development, they appear to be divided into three ranks, each one distinct from the other as regards its importance, capability, and prospects.

In the first rank are the great trunk lines, which, at home or abroad, pass through thickly populated districts, rich in manufactures, minerals, or shipping industries, with their enormous movement of materials and people, and consequently requiring the most ample works, equipment, and appliances for security.

In the second rank may be classed those railways which run through ranges of country where the population is moderate, or where the manufacturing industries are few in number and of minor importance. Although of the utmost value to the community of the long series of small towns and agricultural districts through which they pass, and forming the only great commercial highway, or connecting link, with some distant seaport, or leading business centre, the traffic returns upon such lines are too small to permit of the introduction of the more complete appliances and luxuries to be met with on the richer railways. In newly opened-out countries, and in distant colonies, such lines have often to struggle on for years against financial returns so small as to barely enable them to maintain a condition of efficiency; but where there are natural advantages in soil and climate, combined with a judicious development of all the available resources, the result will be the raising of the standard of the railway itself, and the enrichment of the entire district through which it passes. When laying out lines of this description, it may be necessary to curtail as much as possible the expenditure on works and equipment, but there should be no hesitation in obtaining liberal quantities of land for future enlargement of stations, or for constructing additional stations on promising sites. The value of the land may be small in the outset, but will be enhanced enormously as the benefits of the undertaking become appreciated.

In the third rank may be grouped those branch lines which, starting from a main passenger or goods line, are laid down to some outlying town, seaport, or mining centre, which, although small, is considered of sufficient importance to be brought into railway communication. In general, these lines are laid to the same gauge as the line with which they connect, and the transfer of merchandise waggons is readily effected at the point of junction. Others, from motives of economy, have been laid down to a narrow gauge, involving the transhipment of all goods and cattle at the station where the break of gauge takes place. Most of these branch lines are laid out through the open country, like an ordinary standard railway, but with a minimum of works and appliances. Others are laid down partly on level public roads, and partly through the fields, and are in consequence subject to a statutory low rate of speed when travelling over those portions on the public roads.

In many cases the construction of second and third rank railways, both at home and abroad, has been largely assisted by state or provincial aid. Such assistance must always be valuable to poor or undeveloped districts, but judgment should be exercised so as not to encourage the introduction of any scheme which would interfere or become competitive with any existing undertaking constructed by public enterprise. So long as capitalists invest their money more from commercial motives than from feelings of philanthropy, it would, to say the least, be unjust and impolitic for any country to adopt a course of competition by national funds, and so check the flow of public money into public undertakings. Ordinary public commercial competition may be business, as each party can value and compare their own prospects; but the competition of a scheme enjoying national aid and free money grants is very apt to become one-sided.

There is every indication that even what may be termed a fourth-rank type of railway is destined to play a very important part in the industrial enterprises of many countries, and that in the form of little lines, made to any convenient gauge, and laid either along public roads or open country, or both, the produce from isolated manufactories, forests, quarries, and large farms will be conveyed to the nearest railway stations with greater facility and at much less expense than by carting along the public highway. Such little lines are available in places where the most sanguine promoter would hesitate to suggest an ordinary railway, and may be found to supply what is felt to be the missing link in the economical transport of a long list of materials of everyday use. As they would be almost exclusively intended for merchandise purposes, the statutory requirements would be on the most moderate scale, and as they would be generally constructed at the cost of the parties who had to operate them, the outlay would be restricted to the actual works necessary for convenience and efficiency. Similar little lines have been in use for many years in the busy yards of large ironworks, shipbuilders, and many other localities, where weighty masses of materials have to be moved from place to place in the course of manufacture, and it would be merely carrying out the same idea to a more extended range. The principal inducement for their introduction is the great advantage, both in convenience and cost, that is obtained by hauling a ton of materials over a pair of rails as compared with carting the same weight along an ordinary road; and as the fact becomes more and more proved by experience, these little fourth-rank lines will become more general. Numbers of them are in use at the present time, and some of them, even of only 2-feet gauge, are doing good service, the little trucks conveying manufactured goods to the nearest railway station and returning loaded with coals and other materials. By making suitable arrangements for passing places and junctions, the system could be carried out to considerable distances in thinly populated districts, and be made available by means of local sidings, to several places along the route. With a narrow-gauge type there would, of course, always be the time and expense of transhipment to or from the ordinary railway trucks in the same way as with the road carts, but the time and expense may be lessened by so constructing the little narrow-gauge trucks that the bodies may be readily lifted off the frames and wheels, and be placed like packing-cases in the railway waggons.

It is natural to look to the railways of the first rank for the latest advances in construction, appliances, and equipment, and it is generally there they are found. Great trunk lines, crowded with traffic of all kinds, have not only the opportunity and means, but all the strong inducements to try or adopt any arrangement which promises greater facilities for dealing with the ever-increasing demands made on their carrying powers.

Passenger and goods traffic are so dissimilar in their requirements that when both of them are steadily increasing it becomes difficult, if not impossible, to work the two classes over an ordinary double line. In some cases much assistance has been obtained by shortening the lengths of the working sections and introducing intermediate electric telegraph block stations between the ordinary stations. Long refuge-sidings have also been introduced at many of the signal-cabins or stations, into which goods trains can be shunted out of the way to allow fast passenger trains to pass through without stopping. Up to a certain extent this arrangement works fairly well, but where there is a very frequent service of fast and slow passenger trains, combined with a heavy and constant service of goods and mineral trains, the two lines of way are practically incapable of accommodating such a number of mixed trains without causing serious detentions. The goods trains must shunt out of the way some time before a passenger train is due, and this frequent shunting into sidings results in hours of delay in the transit of the goods and cattle traffic; and when one of such trains is allowed to proceed again on its way up to another station, dove-tailed as it may be between two fast passenger trains, there is always the tendency to run at a much higher rate of speed than is prudent for the class of rolling-stock of which the goods train is composed. To overcome this difficulty some railways have introduced additional UP and DOWN lines on the busiest part of their system, making four lines of way in all, two of these being reserved for the fast passenger and through trains, and the other two for slow trains, goods, and mineral trains. This arrangement of the four lines has afforded great relief to the traffic of all kinds, and has enabled the service to be worked with much greater facility and punctuality. The goods trains being restricted to their own separate lines, can proceed regularly in their order, at their uniform working speed, without having to resort to the spasmodic fast running too often expected from them when passing over some parts of an ordinary double line. Doubtless this four-line system, or rather the principle of laying down two additional lines of way, will go on extending, and will be accelerated in its accomplishment by the growing demand for still higher speed of our fast passenger trains, and still longer distances to be traversed without stopping. High-speed long-distance through trains can only perform their journeys with punctuality, when the route is kept clear of all other trains or obstructions which might interfere with their free running. Any check or stoppage in their course would cause loss of time and prestige.

It is to be regretted that in so many of the cases where two additional lines of way have been laid down, more space was not left between the sets of rails for the fast traffic and those for the slow. In many instances the dividing space is not more than 7 or 8 feet. It would have been better and safer if it could have been made 20 feet. An ordinary goods train is made up of several kinds of trucks, some empty, some loaded, many of them unequally loaded, all of them subject to heavy work and rough handling, and more likely to give trouble than the higher class vehicle, the passenger carriage. The breaking down or derailment of one or two goods trucks on a line of rails close alongside the fast passenger rails, would in all probability so foul and obstruct the passenger line as to cause a very serious accident to an express train which could not be stopped in time. The greater width would not only provide more clearance in case of breakdowns, but would afford increased safety to the platelayers and other workmen engaged on the line. The permanent-way men have to be very watchful to keep out of danger on an ordinary busy double line, but they must be very much more on the alert where there are four lines of way close together side by side.

In the neighbourhood of large cities and important manufacturing centres, railways have created a distinct traffic for themselves by providing means for a large portion of the population to reside in convenient suburbs. Local trains running at suitable business hours have induced people of all classes to select homes a few miles away from town, and the gradual growth of this suburban traffic has produced its own advantages and requirements. At the large terminal stations platform after platform has been added to accommodate the increased number of trains which arrive in the busy parts of the morning or depart in the evening. Every facility has to be provided to permit of the expeditious ingress and egress of the large crowds forming the respective trains—ample platforms, over-line foot-bridges, subways, convenient booking-offices, waiting-rooms, and left-luggage rooms.

The enormous train service on some of these first-rank lines demands the highest efficiency in the signalling and interlocking arrangements, and the use of any devices which will ensure increased facility and safety in the working of the traffic. With a crowd of trains passing a signal-cabin in both directions, and often over four lines of way, it is quite possible for a signalman to make a mistake which cannot be rectified in time to prevent an accident. To obtain increased security many railways have adopted the lock and block system previously described, or some adaptation of the same principle, and this method of working will go on extending as the traffic increases. These additional appliances entail additional care and inspection, for although automatical machinery may be exempt from the human frailty of preoccupation of mind or forgetfulness, it is somewhat delicate in its organization, and requires constant supervision to maintain its efficiency.

On many of the large lines, much has been done to give improved carriage accommodation. Carriages have been made longer, easier on the road, loftier, better furnished, and better lighted; but there is still a very great deficiency in those conveniences so essentially necessary, especially on trains running long distances without stopping. Drawing-room cars and dining-room cars are no doubt attractive, and may contribute considerably to the popularity of certain routes; but it is questionable whether many of the lines at home and abroad which have adopted such luxuries, have not in doing so commenced at the wrong end, and whether it would not have been more to the public satisfaction to have begun by first providing those conveniences which are found in every carriage on every line in the United States. It is satisfactory to find that there is a steadily growing tendency to so construct passenger carriages that their occupants may, by passages or corridors, communicate with all parts of the same carriage or with the adjoining carriages; and there is every reason to assume that the carriage of the future, either by legislation or consent, will combine both the items of conveniences and intercommunication, and will confer not only greater comfort to the passengers, but also increased protection against those outrages which, unfortunately, too frequently occur under the system of isolated compartments.

It will be instructive to watch the results of the passenger receipts on those lines where only first and third-class carriages are used. The elimination of the second class may at first sight appear an innovation; but if there is not any pecuniary loss sustained, there must be a gain in the reduction of unoccupied seats to be hauled. It is customary to provide in every train a liberal number of spare seats of each class to meet contingencies; and the omission of one class may mean the saving of two or three carriages—a very important item in locomotive power.

On important through lines high-speed running has become a leading feature, and compels a very efficient standard of perfection in works and rolling-stock to effect its attainment. There is no indication of remaining contented with what has been already accomplished; on the contrary, the spirit of restlessness is always urging to do something more. The travelling public speak as calmly now of a speed of seventy miles an hour as they did of thirty-five a few years ago; they thoroughly recognize the value of railways, and they merely desire to travel still faster. The incentives of emulation and competition are ever present to encourage further and further reduction of the running time, and the railway that offers a special fast through service for some of its passenger and mail trains, reasonably expects its popularity and patronage to be in the ascendant. Much has been done in permanent way and equipment to make the present high speeds possible, but more will be required if the speeds are to go on increasing. The passenger carriages for such work must be very substantial, and naturally heavy. The locomotives to haul a long train must have increased power and weight, and will necessitate stronger rails to carry the greater rolling loads. With the present system of motive-power, the heaviest item is the locomotive, and its weight must always determine and regulate the character of the works and permanent way. Rails weighing 90 pounds per yard are becoming common, and there is clear indication that before very long sections weighing from 100 to 120 pounds, or more, per yard will be brought into use on many lines. There will be no difficulty in making a permanent way strong enough for rolling loads very far in excess of anything in the present practice; but it will be costly, and the extra expense per mile, extended over a few hundred miles, will represent a sum so large as to raise the question in many cases whether the probable advantages and additional remuneration to be obtained will warrant the outlay.

To some extent the increased speed may be attained by dividing the present long trains into two shorter trains, with a fair interval of time between them. There are many splendid locomotives now running, which on a fairly level line can reach a speed of considerably over seventy miles an hour with a short train, but would be quite incapable of doing so with a long train. At the same time it is possible that if passengers increase in the same proportion as the inducements provided, the short train might not be sufficient for the numbers presented, and there would be no other alternative but to resort to still greater rolling loads and stronger hauling power.

Perhaps electricity, which has already achieved so many marvels, is destined to take a still more prominent part as a motive-power in the working of ordinary railways, and may help out of the difficulty by inaugurating still higher speeds without the necessity of incurring stronger works or heavier permanent way. In addition to its success in the telegraph, in the telephone, and in its brilliant light, electricity is every day coming more and more to the front as a motive-power. At present many tramways and short lines, some of them in tunnel, some above ground, and many of them with very steep gradients, are successfully worked by electricity; but these, being of modern construction, were specially designed and equipped for that method of working, and none of them as yet resort to high speeds. Such rapid strides have, however, been already made in the progress of this system of haulage, as to promise that both increased power and speed will be forthcoming when the demand for them is made manifest. Various modes of application are being tried: overhead wires, underground wires, conductors on the level with the rails, storage batteries or accumulators, and self-contained electric motors, each and all of them being carefully tested to ascertain the comparative cost and efficiency. Much will depend upon the localities and advantages to be obtained for the respective generating stations. In places where a large, constant, and unutilized water supply is available, a great saving may be effected in the most expensive item of electric working, but in the greater number of cases steam-power will have to be adopted for driving the generating machinery. The main question will be whether electricity in its most approved form of application can haul a ton of paying load for one mile at a less average cost, and at as great or greater speed than the ordinary locomotive. Until there is very clear evidence that electricity is cheaper, there will not be any great inducement for its general use as a motive-power on ordinary railways.

Experiments have been made on some existing railways to ascertain how far this new motive-power can be made serviceable under special circumstances. In one case, a powerful electric motor-car has been introduced for working frequent and heavy trains through a long tunnel, where the atmosphere with ordinary steam locomotives became foul almost to suffocation, and the result has shown that the traffic can be hauled efficiently by electricity, and the air in the tunnel maintained pure and clear. In this instance, the question of cost was of secondary importance, the primary object being to avoid the asphyxiating gases emitted from the ordinary locomotives.

In other cases, specially designed electric motor-cars have been constructed with a view to obtain a higher speed for passenger trains than is at present attained with the locomotives, and the trials made have proved that these cars could reach a high speed, but so far only with limited loads. Experiments are still going on with larger and improved machines, from which it is expected to obtain both high speed and much increased hauling power.

It is more than probable that amongst the earliest practical applications of electric motive-power on existing railways will be its introduction as an auxiliary on the steep gradients of some of the mountain railways abroad. In many of these regions there are millions of gallons of water running to waste down the ravines, a portion of which could be utilized in working powerful generating plant, to drive strong electric motor-cars for assisting the ordinary locomotives up the steep inclines. In such localities, with free water-power, the cost of the electricity would be at a minimum, while the cost of the ordinary locomotive would be at a maximum.

In whatever form the electric motor-car may be designed, we are brought face to face with the old axiom, that there must be a certain amount of weight to obtain a certain amount of adhesion; but there will be one important point in favour of the motor-car, that whereas in the ordinary locomotive the weight for traction can only be distributed over a few working wheels, the electric arrangement may distribute it over a much greater number, and so diminish the insistent weight of each wheel upon the rails. There would also be the saving of the dead weight of the tender, the fuel, water, and other minor accessories, as well as the advantage that the active power would be applied in a rotary form instead of reciprocating.

There are important interests at stake in the perfecting of this new system of haulage, and day by day new developments are being made to add to its efficiency and reduce its cost. Existing railways will, however, naturally require some very convincing proof of the all-round superiority of electricity before adopting that power generally in place of their present locomotives. The latter, with their corresponding workshops and appliances, represent so large an amount of invested capital, as to demand most thorough trials and investigation of the new power before they are superseded; nevertheless, if further experience proves that electrical power is better and cheaper than the ordinary steam locomotives, then the change will undoubtedly be made.

Under whatever system of haulage the acceleration of trains be obtained, the increased speed will call for increased precautions in the selection and proving of the materials to be used in such service. Rails must be made more uniform in quality, and must be free from the imputation of fracture under regular wear. Notwithstanding the great improvements made in the preparation of the steel, and in the rolling, there are still far too many steel rails which break under traffic to allow rail-makers to rest satisfied with their work. Something is still wanting in the manufacture to effectually remove this disposition to fracture. The safe rail, the rail of the future, must be one that may bend and may wear, but will never break under ordinary use in the road. Axles must be stronger and tougher, as they will have to bear greater torsional strains than are now imposed upon them; and the wheels, of whatever type they are made, must be incapable of collapsing or falling to pieces upon the sudden and severe application of the brake-blocks. A train, rushing along at a speed of 70 or 80 miles an hour, may on an emergency have to be brought to a stand in the shortest distance possible, and the failure of either axles or wheels in the endeavour to avert one form of accident would inevitably initiate another.

To permit of unchecked high-speed running, many sharp curves will have to be flattened, bridges will have to be built at busy level crossings; and points, crossings, and junctions on the main lines will have to be reduced to the smallest possible number.

It would be difficult to form an opinion as to how far passenger traffic will go on expanding, but if it continues to increase at the same rate as at present, some railways may find it expedient, and even absolutely necessary, to construct new lines altogether separate and apart from the existing routes, and for the sole use of their fast through traffic. As roadside or intermediate traffic would not form any part of the scheme, such lines could be laid out so as to keep away from the populous districts, where property would be costly, and pass instead through those parts of the open country where the most direct course and easiest gradients could be obtained. Stations would only be required at the very large and important places, and at long distances from each other. Lines of this description, reserved for through traffic only, taken alone, might not pay, but taken in conjunction with the existing lines, of which they would form a part, they might prove to be the best solution of the problem of dealing with a crowded train service, the remunerative earnings of which, placed together, might yield a rich return over the entire system. A project for a separate through line might at first appear a little startling, but we have well-known precedents in the vast expenditure already incurred in the constructing of enormous viaducts and connecting lines to avoid long detours on certain through routes. The widening out of an ordinary double line into a four-line road was at first considered as a rather venturesome departure; and it must always be costly because, in addition to the earthworks and permanent way, there is the doubling of all the over and under bridges and waterways, besides the great and expensive alterations at stations. Practically it is almost like making a second railway, and yet the constant extension of the principle is an admission that the working results have proved satisfactory, in spite of the large outlay. A little later the question will force itself more prominently into notice, whether the four-line track or the separate fast through traffic lines, will best answer the purpose. The former possesses certain advantages, but the latter would give more freedom for high-speed running.

Engineers have brought railways to their present stage of perfection, and the public will expect them to devise and carry out still further improvements as the march of development moves onward. It is a simple matter to arrange the traffic on a railway when all the works and appliances are appropriate for the service to be performed; but the advances which are made follow one another so rapidly as to necessitate constant study and organization to effect the structural alterations and additions requisite to maintain an up-to-date standard of efficiency. The traffic manager on a railway receives his instructions from the directors or controllers of the company as to the working out of the train service, rates, charges, and other items of his department, but the engineer has to stand alone, and his technical knowledge and professional skill must enable him not only to design and construct works suitable in character, extent, and strength to the duty for which they are intended, but also to decide when structures are no longer capable of properly sustaining the increasing loads brought upon them, and must be taken down and replaced with others of a stronger description. For this reason the engineer must carefully consider every circumstance and local feature which may influence the design to be prepared; he must thoroughly investigate the nature of the ground for foundations, as the description when ascertained will frequently determine the class of work to be erected, whether in viaducts, bridges, or buildings; and in his selection of materials and calculations for strength, he must allow ample margin to meet further increased weights, as well as for natural deterioration.

He should, indeed, go a little further, and as his perceptive ability and training will always enable him the more readily to foreshadow the direction in which improvements or changes are tending, he should study out and be prepared with his schemes to meet the new departures as the requirements gradually arise.

Strength and efficiency are the leading points which must be always kept in view, and the engineer must never forget that he is solely responsible for the safety of the line and works, and of the public passing over the same.